Where the Germs Are

A SCIENTIFIC SAFARI

Nicholas Bakalar

John Wiley & Sons, Inc.

0-471-15589-6
ISBN 9781684428762 (paperback)
ISBN 9780471155898 (hardcover)

Printed in the United States of America

For John F. Thornton

Contents

Preface

Dirty! Don't touch! Yuck! Feh! You don't know where it's been! These admonitions ring in our ears, for some of us our earliest memories of parental exhortation, for others the indelible mark of our deepest fears. Germs, as we know, are everywhere, lying in wait to attack the inadequately vigilant or insufficiently armed, gangs of serial killers on a random search for their next victim. We do not mock. Well, maybe we mock a little, but in fact, Mother (or Father) does sometimes know best—some germs can be very nasty invaders indeed. Yet we live in a world of microbes, some dreadful, some harmless, some essential to our continued life on earth. Knowing which to avoid, which to eliminate, and which just to live with happily can turn fearful warnings into reasoned discourse and trembling terror into intelligent action. Reassuring? Perhaps. But be warned: there are things out there you're not afraid of that you really should fear, and we will not shrink from pointing them out. Germs, in fact, are almost everywhere you thought they were, and in a lot of places you thought they weren't. This book is a Baedeker for a microscopic country, the land of the most common microbes we encounter in everyday life. It reveals what microbiologists have learned about these creatures (or at least some of what they've learned), and shows the reader how to make his way in a world that is, no matter what we do or how we act, completely infested with invisible beings.

Acknowledgments

I would like to thank the following people for valuable help, suggestions, and advice: James Bakalar, J.D., Kenneth Bakalar, Alan Berkman, M.D., Francine Cournos, M.D., Donald Aislee Henderson, M.D., and Jerry Walsky, Pharm. D. Marguerite Mayers, M.D., read the manuscript with great care, saving me from myself on several occasions and making many suggestions for improvement, every one of which has been incorporated into the text. Isela Puello helped immensely in tracking down references. Wiley's Jeff Golick is a sharp-eyed young editor of the old school, who, I have learned to my delight, brooks no nonsense, syntactic or otherwise. This book would not exist without the ministrations, from its initial idea to its publication, of John F. Thornton, gentleman and gentle scholar, who has honored me with his friendship for more than a quarter of a century.

1

Microbes

AN INTRODUCTION

There was a time, a little more than a hundred years ago, when people didn't believe in germs. Nineteenth-century doctors knew a lot about disease—they could diagnose your illness correctly and tell you quite accurately how much time would pass before you would either get better or drop dead. But because knowledge of the causes of illness was primitive and knowledge of germs was nonexistent, doctors at the time couldn't do much to cure disease or affect its course. Then, in the latter half of the nineteenth century, it became clear that microscopic or invisible organisms caused disease—we've all heard about Louis Pasteur and the other pioneers of microbiology—and that you could prevent disease if you could somehow eliminate the organism that caused it. Remove the handle from the pump in the public square, and you get cholera under control. Heat up the milk to a certain temperature before you drink it, and you can kill the organisms in it that cause illness.

In 1900, the three leading causes of death in the United States were all infectious illnesses: pneumonia, tuberculosis, and enteritis. Along with diphtheria (the tenth most common cause), these diseases caused more than one-third of all deaths. Of these deaths, 40 percent were children under five years old. In 1997, infectious disease accounted for only 4.5 percent of deaths in the United States. The two leading causes of death in 1997 were heart disease and cancer, accounting for more than half of all deaths. Stroke and chronic lung disease were the third- and fourth-ranked killers.

You might conclude from these promising statistics that infectious disease has been gradually conquered by human ingenuity, but you would be only partly right. In fact, in the midst of this hundred-year record of increasing control of infection occurred the worst infectious outbreak in history: the influenza pandemic of 1918 that killed 20 million people worldwide and 500,000 in the United States in the course of less than one year. This is more people than have ever died in so short a period in any war, famine, or natural disaster in the history of the world. Human immunodeficiency virus (HIV) infection, first recognized in 1981, now affects 33 million people worldwide and has caused an estimated 13.9 million deaths. Infectious disease, in other words, is always a lurking threat, however great our scientific progress in battling it.[*]

The effectiveness of antibiotics against bacteria and of vaccines against scourges such as smallpox, polio, measles, and influenza is well-known. And of course these medicines have been immensely powerful, sometimes seemingly miraculous, in curing illness and preventing its spread. But it is less widely known that most of the progress against infectious disease in the United States in the twentieth century was realized before there were any effective antimicrobial drugs at all. Nineteenth-century industrialization caused the U.S. population to shift from rural areas to cities. The result was overcrowding, poor housing, inadequate water supply, and poor waste disposal, which led to vast increases in infectious illness—particularly tuberculosis, cholera, typhoid fever, yellow fever, and malaria. But by 1900, the incidence of many of these

[*]There has been a lot of press about the vast increases in infectious disease over the last decade, but the statistics cited (e.g., a startling report in the *Journal of the American Medical Association* that between 1980 and 1992 the rate of death from infectious disease had increased 58 percent) need some explication. First, those years were the time when HIV rates vastly increased. If you take HIV out of the equation, the increase is only 22 percent. After AIDS, respiratory infections are the next leading cause of infectious disease death during this period, but these deaths are mainly in older people—that is, people who thanks to modern medicine have not died of heart disease, stroke, or cancer, and have lived long enough to die of something else. Death rates in the United States from other infectious diseases—tuberculosis, for example—actually declined during this period, and new AIDS cases have been declining in the United States since 1995. While there is of course reason to be concerned, as is amply demonstrated in this book, there is no reason to engage in fear mongering.

diseases had begun to decline as local, state, and federal governments instituted vast improvements in water supply, sewage facilities, pest control, food safety, and public education. All but five of the 45 states had established public health departments by the turn of the twentieth century. The idea of public health as a governmental responsibility had taken hold, and its success was spectacular. In 1900, 194 out of every 100,000 Americans died of tuberculosis. By 1940 the rate had dropped by more than 75 percent to 46 per 100,000. Malaria had been reduced to insignificant levels thanks to mosquito-control programs. Rat-control measures had made plague all but disappear. Chlorination, begun in the early 1900s, had vastly reduced the incidence of waterborne diseases such as dysentery and cholera. And all this had been accomplished without the use of any effective antimicrobial medicines, because there were none.

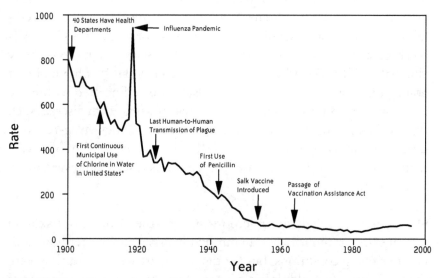

*American Water Works Association. "Water chlorination principles and practices." *AWWA* annual M20. Denver: American Water Works Association, 1973.

Figure 1 Crude death rate, per 100,000 population per year, for infectious diseases— United States, 1900–1996. Adapted from G.L. Armstrong, L.A. Conn, R.W. Pinner. "Trends in infectious disease mortality in the United States during the 20th century." *JAMA* 1999:202:61–6. *U.S. Department of Health & Human Services.*

When antibiotics were developed, they of course were helpful in reducing disease even further, and especially in treating it once it occurred, even though much of the work of controlling infectious illness had already been done. Importantly, antibiotics reduced the occurrence of infectious epidemics. As you can see from the figure on page 3, between 1900 and 1940, every time an infectious disease outbreak occurred, deaths rose, sometimes dramatically. Toward the end of World War II, there was a plateau. Since that time, fewer outbreaks have occurred—but they have not been eliminated.

While an arsenic treatment for syphilis had been discovered in 1909 (Ehrlich's famous "magic bullet") and the sulfonamide drugs in use in the 1930s and early 1940s were to some extent useful in treating infections, the introduction of penicillin during World War II provided the first truly effective chemotherapeutic agent—that is, a medicine that can efficiently kill microbes without also killing the person who takes the medicine. And after the war, along with the discovery of new antibiotics, came the rapid development of vaccines against diphtheria, tetanus, pertussis (whooping cough), and polio. Vaccination programs in the United States and around the world were so successful that the idea of "disease eradication" was born, and in 1977 smallpox became the first (and so far the only) disease ever to be completely eliminated. The World Health Organization (WHO) is hopeful that polio, guinea worm disease, and leprosy can be eliminated within the next few years. But the emergence of previously unknown diseases such as HIV, Ebola, and Marburg and the appearance of new drug-resistant strains of familiar germs means that despite the availability of a wide range of drugs that treat disease, prevention is just as important today as it was in 1900.

Our bodies are filled with microorganisms—they live on our skin, in our guts, in our mouths, and in our bodily organs no matter how sanitary our living conditions, but for the most part we live healthily with them. In fact, we couldn't live without some of them, such as the ones that help us digest food. Much of the protection we have against harmful germs comes naturally, from our own immune systems. What makes a given microbe "harmless"? Only the fact that we have natural defenses that prevent such germs from overwhelming us. One way in which we naturally fight disease is with the phagocytes, macrophages, and more than a half dozen different types of leukocytes (white blood cells) in our

blood. These are cells that ingest bacteria and other foreign particles that find their way into the bloodstream. But we are also protected by a different mechanism in areas of our bodies that have little blood flow. The conjunctivae of our eyes and the membranes of our nasal passages and respiratory systems, for example, are constantly exposed to the thousands of microscopic organisms that are always in the air, yet most of the time we suffer no infection from them. Why not? Because we (like many other animals and plants) produce substances that are natural anti-infectives.

One day in 1921, an English bacteriologist happened to have a cold, so he added a bit of his own nasal mucus to a petri dish just to see what might be cultured out of it. A few weeks later, he noticed that the bacteria growing in the dish—a harmless type of coccus—had failed to grow in the area near the mucus. Something in the mucus was dissolving and killing the bacteria. The bacteriologist called that something "lysozyme," and over the ensuing years of intensive investigation of the substance, he found it in tears; sweat; saliva; the mucus linings of the cheeks; fingernail parings; hair; sperm; mother's milk; the leukocytes and phagocytes of blood; the fibrin that forms scabs over wounds; the slime of earthworms; the leaves and stalks of numerous plants including buttercups, peonies, nettles, tulips, and turnips; and in very high concentrations in egg whites. He had stumbled upon the first natural anti-infective, an enzyme later given the chemical name "mucopeptide glucohydrolase." This scientist would, eight years later, accidentally find something else in one of his petri dishes, a substance that would change the life of almost everyone on the planet. The bacteriologist's name was Alexander Fleming, and he would name this new discovery "penicillin."

Of course, the discovery of penicillin and the many other antibiotics (more than a hundred are in use today) was not the end of the story. Microbes did not succumb so easily to human ingenuity. The slight rise detectable in figure 1 on page 3 over on the far right side represents a resurgence of infectious disease beginning about 1980 that continues to this day. Germs, in other words, are still very much to be feared. In fact, infectious disease is back—in a big and dangerous way. The death rate from infectious disease in the United States today has risen back to approximately where it was 40 years ago. What happened to cause what is apparently the reversal of a nearly century-long downward trend?

Germs reproduce quickly, creating many generations within hours. With such rapid reproduction comes ample opportunity for genetic mutation. And one of the ways germs fight back is by producing genetic mutations that give them resistance to the antibiotics we use to try to eradicate them. Every time we take an antibiotic, we are killing the weakest germs and allowing the strongest—the resistant ones—to survive and reproduce. Eventually, only resistant germs survive, and the antibiotic that was once effective against them becomes less effective or even useless. This phenomenon was noticed very early on in the development of antibiotics. In 1945, it took a total of about 40,000 units of penicillin to cure a case of pneumococcal pneumonia. Today, because the germ is now resistant to low doses, as many as 24 million units a day are given to effect a cure in severe cases. Some diseases for which penicillin was once effective are now completely resistant to it, even in large doses. As new antibiotics are developed, more resistant strains of bacteria almost immediately follow. Vancomycin was until recently the antibiotic used when nothing else would work, but vancomycin-resistant *Staphylococcus aureus* was found in Japan in 1996 and in the United States in 1997. Streptomycin, discovered in 1944, was highly effective against tuberculosis, reducing its incidence from 39.9 deaths per 100,000 in 1945 to 9.1 in 1955. But its usefulness is now limited by the emergence of resistant strains of *Mycobacterium tuberculosis*, and today the disease must be treated with a combination of therapeutic agents. And even these combinations, which can involve more than a half dozen different drugs, fail to work in some particularly bad cases called multidrug-resistant TB.

Yet antibiotic resistance (about which we will have more to say in the course of our narrative), while it is an important reason for the resurgence of infectious disease, is not the only one. Advances in technology, which have obvious great benefits, have their dark side as well. Legionnaire's Disease could not have spread so easily without the presence of modern heating and air conditioning systems. HIV and hepatitis C have both been spread through blood transfusions. Centrally processed food infected with salmonellosis or *E. coli* is efficiently passed around the country, and even around the world, with modern distribution techniques. Tourist trade and economic development in parts of the world that harbor animals infected with human-transmissible diseases have given the human population "new" and potentially devastating germs. Lyme disease, caused by infected ticks carried by deer, is now a problem because what was once agricultural land has become

secondary-growth forest, increasing the population of deer, and sub-
urbs have expanded into such areas, bringing people into closer contact
with these animals. Airplanes are quick and efficient international
transporters of infection. And people whose immune systems have been
impaired by modern medical treatment (e.g., organ and bone marrow
transplants, kidney dialysis, or chemotherapy) are more likely to
acquire "opportunistic" infections—that is, infections that are only
troublesome when the immune system is in some way compromised.

Sometimes you read of the "age of dinosaurs" or the "age of reptiles"
or hear people speak of the time we live in now as the "age of mam-
mals." But in fact we are in—and always have been in since the begin-
ning of life on earth—the age of microbes. Microbes were the first
kind of life, and they remain the most common by far. Bacteria existed
more than three and a half billion years ago, and we have the fossils
to prove it.

No other species, plant or animal, has been as successful. Microbes
live in air and in water, and even where there is no air and no water.
Some bacteria, like the ones that cause food poisoning and those that
live deep under the ocean in sedimentary rock, are called "anaerobic"—
they can't survive if exposed to gaseous oxygen. Others, called "aero-
bic," require oxygen to survive. And then there are some, the facultative
anaerobics, whose attitude toward oxygen is that they can take it or leave
it. They feed on a vast variety of substances—decaying organic matter,
oil, rocks, you name it—and some even make their own food by trans-
forming carbon dioxide into nourishment with the aid of sunlight. They
live in the ice of Antarctica and in the hottest deserts of Africa. They live
in the ocean from its surface to its greatest depths. There are microbes
called archaea (quite different biologically from bacteria or viruses) that
live deep in the ocean under tremendous pressure and near volcanic
vents at temperatures well above boiling. There are microbes that
invade your body and then, once inside, take up permanent residence.
The *Varicella* virus that causes chicken pox is one; the herpes virus is
another. The specialization of their habitats can be remarkable: there is
one species of bacteria, in a genus called *Cristispira*, that lives only in a
certain section of the digestive tract of certain mollusks, and apparently
nowhere else. Various species of bacteria living inside them give cattle
their unhuman ability to use grass as food. Microbes live in your house,
in your bed, in your clothing. They live in every orifice of your body, in

your bloodstream, and inside your bodily organs. They live in vast numbers, far more than any other kind of animal or plant, and there is no place on Earth you can live without them.

When a baby is born, its mouth and digestive tract are sterile. But the first time it feeds, whether on breast milk or cow's milk, bacteria invade the infant, and they will live and thrive inside it for the rest of its life. The human mouth is home to a veritable zoological garden of bacteria. Almost two dozen different species live there—and we're talking about people who are perfectly healthy. Right now—swallow hard—you've got as many as eight different species of *Streptococcus* in there, including, quite possibly, the ones that cause scarlet fever (group A *Streptococcus*), middle ear infections (*Pneumococcus*), and meningitis (*neisseria*), and *Streptococcus fecalis*, which, as its revolting name suggests, also lives in the solid waste matter contained in the intestinal tract. If *Streptococcus fecalis* gets into the bloodstream, it can cause endocarditis (an inflammation of the lining of the heart) in susceptible people. Other oral bacteria, relatively harmless in your mouth, can get into other tissues where they can cause bone, lung, or brain abscesses. Dental procedures can sometimes allow this movement of bacteria from the mouth to other tissue. Dental plaque is the result of an accumulation of various bacteria up to 500 cells thick on the surface of the teeth. Most of these are S*treptococcus sanguis* and *Streptococcus mutans*. The latter produce the lactic acid that causes tooth decay; once lesions are formed, other bacteria help in the process of tooth destruction. The higher the concentration of these bacteria in your saliva, the more cavities you are likely to have. Gingivitis, the kind of serious gum disease that causes destruction of the bones that hold the teeth, is caused by a complex population of many organisms, including *Streptococci* and *Actinomyces*. The actual mechanism by which these bacteria destroy bone is not well understood.

Some of these oral bacteria are probably helpful—they prevent more harmful species from inhabiting the mouth by inducing low levels of antibodies, and some actually contribute to good nutrition by helping to synthesize vitamins.

Staphylococcus lives happily on your skin, up your nose, and in your pharynx. Nontypable *Haemophilus influenzae*, the bacterium that causes ear infections in children, finds a comfortable home in your nose and pharynx as well. *Proteus mirabilis*, a species often found in rotting

meat that probably causes enteritis, can sometimes live in your respiratory system, and so can corynebacteria, which cause diphtheria. The normal vagina usually has about 14 different species of bacteria living in it at any given moment; the normal intestine about 17.

The surface of your eyes—the conjunctivae—are in constant contact with air, and therefore with germs. Fortunately, tears contain lysozymes, those natural anti-infectives, and the constant mechanical washing of the eyes by blinking helps keep them clean. Still, some bacteria are found in the conjunctivae, including *Staphylococcus epidermidis* and some cornyneforms, but usually in very small numbers. The urinary tract enjoys a similar cleansing process because it is flushed with sterile urine every few hours. Yet even here, small numbers of bacteria can be found, probably contaminants from the skin, vulva, or rectum.

If you're beginning to think that there aren't enough antibiotics in the world to clean up this mess, you're probably right. Fortunately, there isn't any reason to clean it up, because in healthy people most of these bacteria are either mildly annoying, completely harmless, or somewhat helpful.

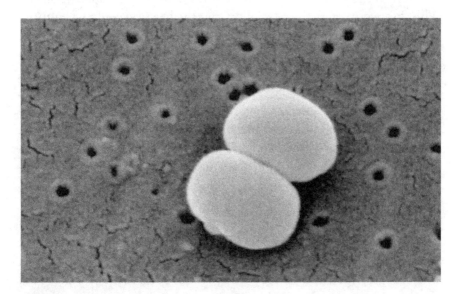

Figure 2 A scanning electron micrograph of *Staphylococcus epidermis*. You can't see them, but you're covered with them and shedding them everywhere you go. They can also commonly be found in human eyes. *CDC. Photograph by Segrid McAllister.*

Bacteria are helpful in the wider world as well. They are at the bottom of the food chain, for one thing. There are bacteria that are essential in breaking down dead organic matter. Scientists have recently discovered that bacteria play a role in species differentiation among certain insects. Different species of wasps, for example, are apparently infected with bacteria that make mating with uninfected species impossible. When the wasps are treated with antibiotics, otherwise incompatible species can interbreed and produce fertile offspring.

Researchers are always finding new ways in which bacteria can be helpful. In December 2000, the journal *Science* published an article written by researchers at the University of Wisconsin and other institutions that showed that certain sulfate-reducing anaerobic bacteria remove the zinc from the waste water produced by mines. These germs are surrounded by beads of zinc sulfide crystals. As they reduce sulfate, freed sulfide ions combine with zinc ions, which then precipitate out of the water. Scientists have known for some time that there are bacteria that can dechlorinate chlorobenzenes, a common industrial compound that can accumulate in the food chain. German scientists, according to a report in the journal *Nature*, have now identified the specific strain. This discovery offers the possibility that the bacteria could be cultured and prove useful in industrial waste cleanup.

Some bacteria are positively delightful, like the ones that make grapes into wine, give yogurt its tang, assure cheeses their multitude of flavors, and lend sourdough bread its pleasant sour taste. Some are normally harmless—our own bodies protect us from them unless we are for some reason susceptible (the very young, the very old, the unhealthy, or those with weakened immune systems). Others are harmful and cause disease even in healthy people. And still others would be harmful if we didn't have drugs to counter their effects.

What is a germ? The question may seem trivial—after all, a germ is obviously a bad microscopic thing that gets in you and makes you sick. Well, yes and no. As we've seen, there are plenty of "germs"—that is, microscopic organisms—that are in you and don't make you sick. And there are organisms that get in you, make you sick for a while, and then, after you get better, just stay in you, sometimes taking up residence in very odd places, as you will see later in this book. For convenience, and so as not to bore everyone to death, I use the term "germ" in this book

rather loosely to describe lots of different kinds of organisms that live in and on us, some of which make us sick, some of which make us sick for a while and then live on without bothering us, and some of which never do any harm at all. The first person who saw a microscopic organism, Antonie van Leeuwenhoek, called them "animalcules." Our term "germs" is no more specific than that term assigned in 1674. The glossary makes some fine distinctions among organisms, but nevertheless we need to clarify a bit here at the outset.

There are somewhere between a half million and a million species of bacteria, and roughly five thousand different viruses. Only a small fraction of these bacteria have been scientifically studied and categorized. Then there is the group of disease-causing one-celled organisms mentioned previously, the archaea, which includes the blue-green algae that constitutes pond scum (and which, confusingly, are not algae at all). There are other organisms that are neither viruses nor bacteria nor archaea that nevertheless infect us and cause disease—some of these we call "parasites," although one could as easily call bacteria and viruses parasites as well. Malaria, for example, is caused by a protozoan called *Plasmodium*. A protozoan is a germ, if you want, but it is a one-celled organism that is a member of the phylum Protozoa, neither a virus nor a bacterium in the classification scheme. Even fungi, which are species of plants, cause disease. *Candida*, for example, is a yeastlike fungus whose various species cause thrush, vaginitis, and some systemic infections. To make things even more complicated, virologists have discovered yet another infectious organism called a "prion," which contains neither DNA nor RNA and is probably composed entirely of a single protein. This doesn't discourage anyone from calling these organisms germs, and it won't discourage us from discussing such bugs in this book.

Bacteria, viruses, archaea, protozoa, fungi, and prions can all be harmful, but they work their mischief in different ways. The distinctions are important because they determine how the diseases they cause can be prevented or treated.

Of the thousands of species of bacteria, most live their lives either with beneficial effects or with no effect at all on humans. But there are a number of them that can make you sick, and some that can kill you. Bacteria are one-celled organisms that reproduce by dividing. There are four groups of bacteria, classified by shape: the *bacilli* (rod-shaped),

the *cocci* (spherical), the *spirilla* (spiral), and the *vibros* (shaped like commas). You may have heard the terms "gram-negative" and "gram-positive." These terms refer to whether or not a given organism can be stained with a certain substance that makes them visible under an electron or phase contrast microscope. (Despite the fact that it is normally styled with a small "g," Hans Christian Gram is the name of the Danish bacteriologist who developed the staining technique in 1844 and has nothing to do with the gram that is a measure of weight.) Gram-negative and gram-positive are one way of classifying bacteria. Penicillin is useful against gram-positive bacteria like *Streptococcus*. It doesn't work against gram-negative bacteria like the one that causes salmonella food poisoning. And these groups are classified further according to the way they live in colonies and their metabolic actions. The taxonomy of bacteria is no less complex than that of any other species—they are divided into kingdom, division, class, order, family, genus, species, and subspecies just like other animals and plants.

In general, all the species of bacteria that make you sick do so in pretty much the same way: they invade your cells, making the body's immune system react in some way, usually by causing tissue inflammation. How they get into your cells varies from one kind of bacterium to another. Campylobacter, for example, a gram-negative genus of which there are several species, can cause infectious diarrhea, appendicitis, and an acute inflammatory nerve disease called Guillain-Barré syndrome, which causes (usually temporary) paralysis. A relatively small number of these cells can cause very serious illness. Campylobacter produces a kind of glue, a protein called adesin, which sticks the germ to the wall of the intestinal mucosa, making invasion of the tissue possible. It then produces toxins that destroy the cells. The group of *E. coli* species that probably produces most cases of "traveler's diarrhea" invades the cell with the help of fibrils, tiny filaments that penetrate parts of the cell's genetic material to gain entrance. Chlamydia, another gram-negative bacterium that causes various symptoms and disorders depending on the species, can reproduce only by invading the tiny empty spaces called vacuoles in the plasma of the cell. There they produce a form called "reticular corpuscles," which are released to invade, cause inflammation, and reproduce in other cells.

Protozoa, responsible for malaria and many other diseases, reproduce and infect in a variety of ways. Cryptosporidium is a common waterborne parasite that is usually harmless to humans, except those

whose immune systems are for some reason compromised. Under the right circumstances, however, the parasite flourishes, and some of them convert into a form called an oocyst, which is very resistant to ordinary water treatment procedures. If this oocyst is ingested by an animal (including humans), it can release sporozoites, which attach to the surface of the intestines to initiate a new cycle of infection that causes abdominal cramping and explosive watery diarrhea. No one knows for sure, but the nature of the diarrhea suggests that the cause is a toxin that poisons the cells of the intestines.

The protozoan that causes malaria has a life cycle that begins when a mosquito of the species *anopheles* feeds on the blood of an infected person. The protozoan (which comes in four different species called *Plasmodium falciparum, P. vivax, P. malariae,* and *P. ovale*) then undergoes sexual development inside the mosquito, leaving sporozoites in the insect's salivary glands. When the insect bites another human, these sporozoites are released into that person's blood. They reproduce asexually in the cells of the liver. After two to four weeks, they are released into the bloodstream in a form called merozoites, which invade and destroy red blood cells. The parasites can persist in the liver (in the case of *P. vivax* and *P. ovale*) or in the blood (in the case of *P. falciparum* and *P. malariae*), where they cause recurrent malaise, fever, headaches, chills, and bodily pain that may persist for months or even years. People who are otherwise in good health usually recover from malaria even without treatment, although early treatment with quinine drugs is very effective. The exception is falciparum malaria, which if left untreated can kill by destroying the cells of the liver, spleen, and brain.

Viruses, unlike bacteria and protozoa, can live and reproduce only inside cells, and they are much smaller than either of those types of organisms. They consist of an outer protein or lipid shell that surrounds a nucleic acid core, which may be either RNA or DNA. And they can infect not only the cells of animals and plants, but the cells of bacteria as well. There are about a hundred of them that infect humans, and they cause diseases as mild as head colds and as deadly as AIDS. Some have long incubation periods and can cause one disease upon initial infection and another after years producing no symptoms at all. Varicella zoster, for example, the virus that causes chicken pox, can be contracted in childhood and then live in your body permanently, causing shingles, a painful nerve disorder, much later in life. And there are many viruses that infect humans without causing any symptoms at all—the only way

you can tell that you're infected is by laboratory tests that reveal antibodies to the virus in your blood.

Viruses have reactive sites on their outer shells that interact with receptor sites on the cells they are attacking. These "key-in-lock" reactions are quite specific—a given virus is limited to infecting certain kinds of cells in certain species of animals. Once the virus finds the right receptor site, it can either fuse with the host cell membrane or move inside the cell. Either way, it can then insert its nucleic acid into the cell and then reproduce—that is, messenger RNA is "transcribed" from viral DNA or RNA. This process can take place in the nucleus of the cell (as it does with the herpes virus, for example), in the cell's cytoplasm (like the polio virus), or at the cell's surface (the flu virus does it this way). The final step is the release of new infectious material from the cell, which can then go on to infect other cells. This happens in various ways depending on the virus, including by the disintegration of the host cell wall. Because viruses replicate in such an intimate way with the cell itself, it is very hard to develop a medicine that will kill the virus without poisoning the host as well. This is why there are so few effective treatments for viral illnesses.

If bacteria, protozoa, and viruses aren't enough to worry about, you can always think about the recently discovered infectious agent called a prion. Bacteria, viruses, and protozoa are clearly living things. With prions, it isn't so clear. They apparently have no nucleic acid, and consist of a protein alone. That a protein can by itself transmit disease was quite surprising to microbiologists, but this nevertheless appears to be the case. Prions produce various diseases in animals and humans that are often referred to as "spongiform encephalopathies." "Spongiform" means just what it looks like it means—having a form that resembles a sponge. "Encephalopathy" refers to diseases of the brain. On postmortem examination, the brains of animals (including humans) infected with these diseases are marked by large empty spaces in the cortex and cerebellum, giving the organ the appearance of a sponge.

Prion diseases all have similar and extremely unpleasant symptoms: loss of motor control, paralysis, dementia, emaciation, and death. There are four forms of the disease in animals, affecting sheep, mink, elk, mule deer, and cows. The "mad cow disease" of recent headlines is bovine spongiform encephalopathy. In humans, this prion causes Creutzfeldt-Jakob disease (CJD), Gerstmann-Straussler-Scheinker

syndrome (GSS), familial fatal insomnia (FFI), and kuru. CJD in humans can be contracted by eating infected beef. Kuru, one of the first prion diseases discovered, was found among isolated tribes in New Guinea where religious ceremonies in which the brains of dead relatives were ingested were identified as the cause of transmission. Prion diseases in infants are called Alpers Syndrome. Somewhat mysteriously, these diseases appear in some cases to be genetically transmitted as well as acquired. No one knows exactly how.

Some organisms are known to release poisons that kill. Certain kinds of algae, for example, that live in seawater can poison fish and shellfish, which then store the poisons in their tissue. Eating one of these infected animals can produce dramatic effects: demoic acid produced by a species of algae called *Nitzschia pungens*, for example, can cause amnesia. Others produce toxins that can paralyze you. Some just give you diarrhea; some can kill you. Some molds and fungi also release substances that cause disease in humans.

Many animals are "vectors" of infectious disease—that is, they carry disease and transmit it to other animals or humans. Mosquitoes, as we saw previously, are a vector of malaria. Birds carry West Nile virus. Cockroaches, mice, rats, flies, and many other animals can cause disease in humans by transmitting infectious organisms. These animals are important because they are the ones with which we humans have a very close relationship—second in intimacy only to our relationship with germs themselves.

Viruses, bacteria, archaea, prions, protozoa, and fungi all exist in nature. Disease does not. Disease is a human invention, a method of characterizing and organizing symptoms, not a phenomenon that exists out there apart from us. From the point of view of humans, HIV infection is a disease, one of the most terrible there is. But from the point of view of a human immunodeficiency virus, HIV infection is merely life. From the point of view of some organisms, human beings themselves are a disease. You might say (even if the analogy is somewhat imperfect) that tigers, for example, have a bad case of "humans," so bad that it may in the end wipe them out completely. We're infested with microbes of many kinds, but we are infected with a disease only when one of those microbes causes troublesome symptoms. And of course even microbes that are ordinarily considered disease-causing don't always cause disease.

Antibiotics are not the only weapons against infectious disease. Several other important technological changes have helped considerably. Serologic testing, first used in the early years of the twentieth century, made diagnosis much more reliable. Syphilis and gonorrhea were widespread diseases at the beginning of the century, but they are difficult to diagnose during latent stages. Serologic testing revealed in 1901 that between 5 percent and 19 percent of men in New York City were infected with syphilis.

Isolating germs in order to determine whether they are disease-causing is an essential technique that first came into use in the late nineteenth century. The first method used was to strain infected material through a series of smaller and smaller screens and then inject the final product into an animal or plant to see if the disease would be produced. This technique was used in 1898 to isolate the organism that causes tobacco mosaic virus, and then in 1900 to discover the virus that causes yellow fever. By the 1930s, techniques had been developed to culture viruses, a technology that allowed large-scale production of live or killed viruses from which vaccines could be developed. Staining techniques to make viruses visible under electron microscopes were widely used by 1960.

Today, techniques for identifying infectious pathogens are even more sophisticated. Nucleic acid hybridization and sequencing (tests that examine the genetic code) have now been used to identify the organisms that cause hepatitis C, human ehrlichiosis, hantavirus pulmonary syndrome, Nipah virus disease, and AIDS. Understanding infection at the level of molecules has led to knowledge of new methods of prevention and treatment. The replication analogs and protease inhibitors used to treat HIV infection, for example, are based on a knowledge of the virus's method of replication at a molecular level, and their success depends on using that knowledge to target certain sites on the virus to limit or eliminate its ability to cause harm.

Now you have some of the basics of what scientists know about microorganisms, what germs do, and how they do it. As you might imagine, the details are much more complicated than what we've outlined here, and so much research is going on in microbiology that what I've said may soon be outdated. In the following chapters we'll try to apply some of this knowledge to the most common and most unavoidable of the germs we live with.

2

Hungry? Let's Eat

THE CONTAMINATED KITCHEN

Rabies is a virus, and the usual route of transmission to humans is by the bite of an infected animal. Rabid dogs are the main transmitter of the illness in Latin America, Africa, and Asia. In the United States, where vaccination of pets has controlled the disease, the infection is uncommon, and when it does occur it is usually because the person has been bitten by a wild animal. Domestic animals, if they aren't vaccinated, can get rabies from wild animal bites, too, and that is apparently what happened to a cow in Worcester County, Massachusetts, in November 1998. Because of a series of events, this presented a big problem.

On November 6, the cow seemed to lose its appetite and began salivating excessively. At first the vet thought it was an intestinal obstruction that was causing the problem, but the cow soon started to lose control of its muscles and act very aggressively—not symptoms of a bellyache. Two days later it was dead, and an examination of its brain tissue revealed that it had been infected with a type of rabies virus usually carried by raccoons in the eastern part of the United States. No one knew when the presumed encounter with the raccoon had taken place, but the cow had been milked 12 times in the week before it died, and the milk had been pooled with milk from other cows and sold. Since pasteurization kills the rabies virus, there wouldn't be any problem with the milk—except that a portion of the milk produced between October 23 and November 8 was sold unpasteurized. Sixty-six consumers of "all-natural" unpasteurized milk got a little more "nature" than they'd

bargained for, and all of them had to undergo postexposure prophylaxis for rabies. Rabies is one of the few diseases that is nearly 100 percent fatal if untreated and nearly 100 percent preventable if treated quickly after exposure. So everyone except the cow survived.

The Massachusetts incident of the rabid cow isn't all that unusual. In fact, since 1990 an average of 150 rabid cattle per year have been reported to the Centers for Disease Control (CDC). The transmission of rabies by eating the meat or drinking the milk of infected animals has never actually been proven—the much more common route of transmission is by contact with infected saliva, and in the cases where transmission by milk or meat has been suspected, the more common route has never been completely eliminated as a possibility. In any case, the rabies virus has been found not only in milk, but also in the kidney, prostate, pancreas, and other tissues and body fluids of animals, and transmission by ingestion is a theoretical possibility. The CDC just doesn't know how high the risk really is. Pasteurization and cooking make the rabies virus harmless, and the CDC recommends that all dairy products be pasteurized before consumption.

So it isn't only wild animal bites you have to worry about when it comes to rabies—you can probably get it in the food you eat. The fact is that when you eat, even when you eat food prepared in your nice clean kitchen, you can be eating a lot of stuff you can't see. Rabies in food is very rare. But other germs are quite common. Most people think they're more at risk for bacterial food poisoning in food prepared outside their homes. One survey found that 17 percent of people attributed food-borne illness to food handling at home, while 65 percent blamed restaurants for the problem. In fact, however, food poisoning incidents in homes are much more common than general outbreaks among people consuming food in public places. You just don't read about them in the newspapers. Your nice clean kitchen, and the things you can't see that live there, are the subject of this chapter.

At the beginning of the twentieth century, things were a lot worse. People didn't understand much about food-borne illness—in fact, it wasn't until about 1916 that people began to realize that diseases like pellagra, rickets, and beriberi were nutritional illnesses, not infectious, and that vitamins ("vital amines") were an important part of good health. But once people realized what was causing food-borne illness, they were rapidly able to do many things—without any antibiotics or other medicines—to prevent them. Refrigeration, pasteurization, hand

washing, sanitation, pesticide application, and public education vastly reduced the incidence of food-borne illness long before the existence of antibiotics or vaccines. Typhoid fever, for example, infected 100 people per 100,000 population in 1900. By 1920, it had dropped to 33.8 per 100,000.

Dr. Charles Gerba, a microbiologist at the University of Arizona, did a survey and found that in most homes the bathroom is much cleaner than the kitchen. Kitchen sinks, he found, are the place where you can find the worst kinds of germs, and thorough wiping of sink and counters with a sponge is an efficient method for spreading them all over the kitchen. This sounds scary, but what Gerba doesn't claim, because it isn't true, is that this makes lots of people ill. Obviously, it doesn't. Most people who use their kitchens every day rarely get infected with anything.

This doesn't mean there aren't illnesses you can catch in there. There certainly are. Food can spoil—that is, germs can flourish in it. And germs can make you sick. Let's start with what is perhaps the best-known food-borne germ of all—salmonella, subject of endless newspaper articles and radio talk show chatter.

Raw Eggs and Dirty Cutting Boards

In 1999, the CDC tracked nine different food-borne illnesses caused by bacteria, and confirmed by laboratory analysis 10,697 cases. Salmonella infection accounted for 4,533 of these, a little more than 42 percent. Salmonellosis is actually caused by a group of bacteria. The most common serotypes in the United States are called *Salmonella enteriditis* and *Salmonella typhimurium*.* The name has nothing to do with fish—it's named after the U.S. scientist who first isolated it, and people figured out more than 100 years ago that it caused illness. Salmonella are gram-negative bacteria that move around with the aid of "flagella," hairlike filaments attached to each cell.

Many people are extremely conscientious about washing off the cutting board on which they cut chicken before placing anything else

*These are actually serotypes, not separate species, but they are usually styled as if they were species names: *S. enteriditis* or *S. typhimurium*. The CDC is stricter: they always use the style *Salmonella* serotype enteriditis.

on the board. They've heard that raw chicken can contain salmonella, and they're right—salmonella can be transmitted this way, but it's not the only way. It is usually transmitted by contact with feces of other people or animals. It can be transmitted, for example, from the hands of a food handler who forgot to wash his hands after he used the toilet. (Of course, the food handler has to be infected himself to transmit the disease.) Since animal feces can also contain the bacteria, edible meat and poultry can be infected in this way, and salmonella from infected raw meat or poultry can move, via hands or cutting board, to vegetables, which, if eaten raw, could cause disease. Pet birds and reptiles can be infected, too, and handling them, as we'll see in chapter 7, can transmit salmonella. Another common source of salmonella is raw or undercooked eggs, because salmonella can infect the ovaries of hens, which then lay eggs containing the bacteria (the hens don't show any symptoms of disease, so you can't tell a healthy one from one whose ovaries are infected). At any given time, only a few hens are infected, and contaminated hens lay mostly uninfected eggs, with only an occasional contaminated one. In the northeastern part of the United States, where the disease is most common, 1 in 10,000 eggs is infected. However, since eggs in commercial or institutional kitchens are often pooled, the bacteria can easily spread to eggs that were uninfected. The CDC estimates that about 1 in 50 consumers per year might be exposed to a contaminated egg—but of course of those exposed only a minority will get sick. Even if you are unlucky enough to have an infected egg in your refrigerator, cooking destroys salmonella, so cooked eggs are harmless. Pasteurization also kills salmonella. There isn't any easy way to tell if an egg is infected—they look and taste the same as uninfected eggs.

What could be healthier than raw sprouts? Sprouts—alfalfa, radish, and clover—are practically a synonym for health food. But since 1995, raw sprouts have emerged as a carrier of salmonella infection. Often the seeds themselves are infected, so no amount of careful handling by you or anyone else will help, and even growing them yourself won't guarantee that they're not contaminated. The Food and Drug Administration has this simple advice for people who want to avoid food poisoning from raw sprouts: don't eat them. Most people will find this a small sacrifice.

Salmonella infection rarely has serious consequences. The number of cases reported to the CDC varies from year to year and averages about 8,000. But the CDC usually gets notification only of widespread

public outbreaks. There are probably about 30,000 more infections per year that are not reported and never even come to the attention of a doctor, since the disease is almost always mild and transitory—a few days of unpleasantness with diarrhea, abdominal cramps, and sometimes a slight fever. Salmonella poisoning kills about 500 people a year, about the same as the number of people who die in car accidents over Thanksgiving weekend. Most of the deaths are among the elderly in nursing homes. Infants and the immunocompromised are also at risk for serious illness.

So the danger of salmonella infection should not be exaggerated, but it shouldn't be minimized, either, and sensible steps should of course be taken against transmitting it. What can you do to avoid infection? First, and most important, wash your hands often, especially after using the toilet and especially while you are preparing food. Eating only pasteurized foods or thoroughly cooking all food will kill any salmonella bacteria in it and render them harmless. But there are some foods you still want to eat raw. Keeping salad vegetables separate from meat is also an easy step to take. If you prepare meat on a cutting board— you've undoubtedly heard this advice before—wash the cutting board with soap and hot water before you cut vegetables on it.

The seriousness of a salmonella infection depends on the number of bacteria you ingest. Since normal refrigerator temperatures prevent the bacteria from reproducing, keeping eggs refrigerated can prevent disease even if you eat an egg that has some salmonella bacteria in it. That goes for foods that contain eggs as well. Salmonella can be transmitted on the shells of eggs, but since the 1970s, procedures for washing and inspecting eggs have made this extremely rare. In any case, you should discard any cracked or dirty eggs. Wash your hands and cooking utensils after contact with raw eggs. When you cook eggs, eat them right away—keeping them warm for more than two hours is not a good idea. Homemade ice cream and eggnog are often made with raw eggs— the only way to avoid a possible infection from these foods is to avoid eating them. Commercial eggnog and ice cream are made with pasteurized eggs, and these products have never been linked to salmonella infection, so the greatest danger in eating them is their high fat content. Although it is probably easier said than done, you should avoid unpasteurized raw eggs in restaurants in things like hollandaise sauce and Caesar salad dressing. Restaurants should, but don't always, use pasteurized eggs in recipes that call for pooling of raw eggs.

Some states require refrigeration of eggs from producer to consumer, but not all do. The U.S. Department of Agriculture now tests breeder flocks that produce egg-laying chickens to make sure they are not infected, and the Food and Drug Administration has issued guidelines for handling eggs in restaurants and other public food preparation facilities.

Slurred Speech and Blurry Vision

On June 30, 1994, a 47-year-old man was admitted to a hospital in Arkansas. He was nauseated and his vision was blurry. His speech was slurred, and he was having difficulty swallowing. His gag reflex was impaired. His eyelids were drooping and his face muscles weren't working properly—he seemed to have some sort of paralysis. When doctors tested his muscles by electromyography—stimulating them with electric currents—they found an incremental response to rapid stimulation. They asked him what he'd eaten, and he reported that he'd had some home-canned string beans and a stew made from roast beef and potatoes. The home-canned string bean report set off alarms—home canning is a notorious breeding ground for a bacterium called *Clostridium botulinum*. If the beans contained that organism, it was definite: the man had botulism. So they examined the beans and were mildly surprised to find no *C. botulinum*. But then they took a look at the stew and found what they were looking for. The stew had been cooked, kept on the stove unrefrigerated for three days, and then eaten without reheating. The covered pot had provided exactly the kind of oxygen-free environment in which *C. botulinum* thrives.

If salmonella is the most talked about form of food poisoning, botulism must come in a close second—in fact, the word is often used as shorthand for "some deadly poison in food." But food-borne botulism is very rare. While 60 outbreaks of salmonella poisoning were reported to the CDC in 1997, there was only one outbreak of botulism, involving two people, both of whom survived. Of course, there are always more unreported cases of disease than reported, and this number does not include the 80 cases of infant botulism reported in that year, a special kind of botulism infection that we'll discuss in chapter 5.

C. botulinum is an anaerobic, gram-positive bacterium that produces a toxin whose effects can be severe or fatal. The name comes from

botulus, Latin for sausage, and is so named because many of the earliest cases in Europe were associated with home-fermented sausage. Sausage is almost never the cause of botulism in the United States, however. The bacterium is contained in spores, which are very resistant to heat. They can survive cooking, even boiling, and then germinate and start producing poison in the food when it cools. Since they don't require oxygen, they can live comfortably inside a can of preserved food. *C. botulinum* is actually a group of seven separate organisms, labeled A through G, which can be distinguished by the type of nerve poison they produce. A, B, E, and (rarely) F cause disease in humans, and C and D affect birds and mammals. Type G, which was identified in 1970, doesn't appear to cause disease in humans or animals. For reasons that no one completely understands, the types are geographically distributed—type A is most common west of the Mississippi, type B in eastern states, and type E in Alaska. The six types that cause disease produce neurotoxins that are quite similar. They do their damage by inhibiting the release of acetylcholine from the nerve's synaptic terminal. This prevents nerves from properly communicating with each other, resulting in muscle malfunction and paralysis.

Alaska's food-borne botulism problem is particularly acute: 27 percent of U.S. botulism cases occur there, and the Alaskan rate is among the highest in the world. Type E, the most common type in Alaska, is associated with fish, both salt- and freshwater varieties. There have been more than 100 outbreaks of botulism in Alaska since 1950, the latest of which occurred in January 2001. All the cases have been associated with eating fermented food, traditionally prepared by Alaskan natives in a grass-lined hole or in a wooden barrel sunk in the ground. Since the 1970s, they've used plastic or glass containers, and the fermentation has been done aboveground or indoors. The preparation by this more "modern" method has made botulism more common—the production of botulism toxin is more likely at warmer temperatures, and rates have increased during the last quarter of the twentieth century. Early diagnosis and antitoxin treatment have helped—there have been no deaths from botulism poisoning in Alaska since 1994. The CDC and the Bristol Bay Area Health Corporation, a health care delivery firm operated by Alaskan natives in southwest Alaska, have undertaken a campaign to educate Alaskans about botulism and to encourage natives to return to traditional ways of fermenting food. In this case, it isn't just nostalgia: the old ways really are the best.

People who do research on botulinum toxins must be very careful and very brave. Even minute quantities of the toxins that that are inhaled, eaten, or absorbed through the eyes or breaks in the skin can cause acute poisoning and death. Laboratory technicians who handle the stuff have to undergo special training, and they have to be inoculated with botulinum toxoid vaccine, a special preparation the CDC makes available to people who work with the substance. Then they dress up in disposable lab coats, surgical gloves, and face shields before they go to work. The benches they work at have special Plexiglas shields. Handling clinical specimens from patients is risky enough; working with cultured or concentrated toxin is perilous.

Like other kinds of food poisoning, you're much more likely to get botulism at home than in a restaurant: between 1950 and 1996, 65.1 percent of botulism cases were traced to home prepared food and 7 percent to food commercially prepared, including in restaurants. The type of processing involved in the remaining 27.9 percent of cases is not known. Being a vegetarian may be good for your health in general, but it's not so good for avoiding botulism, because most cases of botulism are transmitted in vegetables. In second place is seafood—87 of 444 outbreaks during that 47-year period were caused by marine products. Beef, dairy products, pork, and poultry have been associated with a minority of cases.

In many instances, treating food in a way that would completely eliminate the bacterium would make it inedible. To be sure you'd killed the spores, you would have to heat the food to 248°F for at least half an hour. That wouldn't make your vegetable medley taste too good. So avoiding the bacterium or preventing it from producing the neurotoxin is the most successful approach. Sealing food in airtight packages doesn't help—in fact, it encourages the growth of *C. botulinum*, which doesn't like oxygen. A 10 percent solution of sodium chloride—table salt—helps inhibit growth and toxin production, as do food preservatives such as nitrite, sorbic acid, paragens, phenolic antioxidants, and ascorbates. Lactic acid bacteria—the kind that give yogurt its flavor—also inhibit toxin production in *C. botulinum*.

Food isn't the only route of transmission for botulism. There is a form called wound botulism in which, as the name suggests, the bacterium invades and reproduces in an open wound or sore. We'll have more to say about this in chapter 7. In addition, there have been isolated cases of botulism in which the organism appeared to be coloniz-

ing the gastrointestinal tract without any infected food having been ingested. In some cases, the patients had had a history of disease or gastrointestinal surgery that may have predisposed them to colonization, but in other cases, the origin of the bacterium remains a mystery. These cases of undetermined origin are quite rare.

You can get a mild case of botulism that makes you slightly ill with blurred vision, nausea, vomiting, or diarrhea. In such cases, people usually don't seek medical care; the symptoms go away by themselves, and that's the end of it. But if the symptoms persist and you start having problems producing clear speech, difficulty in swallowing, weakness in your extremities, and trouble breathing, then you need emergency medical care.

If botulism is diagnosed, the immediate treatment is to induce vomiting and wash out the stomach to get rid of any poison that hasn't already been absorbed. The worst problem is respiratory failure, and for this reason, anyone who gets a severe case of botulism poisoning has to be hospitalized. Infected patients must be housed in an intensive care unit where intubation, tracheotomy, and mechanical ventilators can be used. There is also an antitoxin that should be administered, but it only prevents further damage and doesn't reverse any damage that has already been done. You have to weigh the benefits of the antitoxin against its risks. The toxin comes from horses, so allergic reactions to it are possible, which can in some instances themselves be deadly. The toxin is helpful if you receive it less than 72 hours after initial symptoms. After that, it doesn't do much good.

Death from botulism is rare, but when it does occur it is either because the disease wasn't recognized early enough or because there were complications from the long-term mechanical respiration treatment. This treatment in many cases has to go on for several weeks, and mechanical breathing aids can sometimes be necessary for up to seven months before normal muscle function returns and the patient can breathe on his own. The man hospitalized for botulism infection in Arkansas recovered, but not before he had spent a full seven weeks in the hospital, including 42 days on mechanical ventilation.

Botulism has never been a common disease, and it is getting rarer and less deadly every year. The 1950s saw an average of 233 cases per year; by the 1990s, the average was 123. The fatality rate has also decreased markedly, from about 25 percent in the 1950s to about 5 percent in the 1990s. This decline in death rate is probably due to improvements in respiratory intensive care and the prompt administration of antitoxin.

There is another species of this bacterium called *C. perfringens* that is more common than *C. botulinum* but causes less serious illness. The symptoms are similar, and there is at least one serotype that can be fatal, but the CDC reports no deaths among the 2,772 cases reported between 1993 and 1997, a period that saw only 56 cases of *C. botulinum* infection.

When Pepto-Bismol Won't Do

More common than *C. botulinum*, and perhaps as notorious, is the bacterium *Escherichia coli*. There are hundreds of different serotypes of this gram-negative bacterium, any one of which can cause disease, and many of them are part of the normal bacterial fauna of the intestines. They can get into other parts of the body and cause trouble—75 percent of urinary tract infections, for example, are caused by *E. coli* that got into the wrong place. Appendicitis almost always involves infection with a strain of *E. coli*, among other organisms. In 1982, researchers discovered a serotype called O157:H7[*] that when ingested releases a toxin that causes direct damage to the mucosal wall of the large intestine. This can result in some real unpleasantness, including severe abdominal cramps and bloody diarrhea that lasts for a week or more—and that's in the mild cases. About 5 percent of cases have life-threatening complications. One is called hemolytic-uremic syndrome (HUS), which consists of destruction of red blood cells (hemolytic anemia), a reduction in blood platelets (thrombocytopenia), and kidney failure, allowing waste products to accumulate in the blood instead of being excreted in the urine. Another is thrombotic thrombocytopenic purpura (TTP), which includes all of the above symptoms plus fever and neurologic problems. These complications occur most frequently in children under 5 and adults over 60.

When *E. coli* makes the news, it's usually because of an outbreak caused by eating or drinking something in a public place. That's what happened on September 3, 1999. Ten children in counties near Albany, New York, were hospitalized with bloody diarrhea, and everyone

[*]The serotypes are categorized according to characteristics of their antigens, the substances that cause the immune system to react: somatic (O), flagellar (H), and capsular (K).

wanted to know why. Health officials quickly discovered two facts: they were all infected with *E. coli* O157:H7 and they had all attended the Washington County Fair, held two weeks earlier. By September 15, more than 900 fair attendees had reported diarrhea, and stool cultures found that 116 of them were infected with *E. coli* O157:H7. Sixty-five people were hospitalized, and eleven children had developed HUS. Two people, a 3-year-old girl and a 79-year-old man, both of whom had HUS, died. Finally, an environmental study of the fairgrounds uncovered the truth: one area at the fair where several food vendors used water to make drinks and ice was supplied by a well. The well water was highly polluted with *E. coli*, and careful testing showed that the strain was O157:H7.

But the cases that don't make the newspapers—the ones that occur in homes—are actually more numerous. The kitchen is a fine place to pick up this bacteria because the main reservoir of the germ is in food, usually in cattle. The germ lurks in undercooked hamburger meat and unpasteurized milk, and since cattle carry the germs without having any symptoms of disease, you can't easily tell which cattle are infected.

Because the germ is shed in human stool, you can pick it up by that route, too. You wouldn't think that ingestion of other people's stool would be a frequent occurrence, but when someone forgets to wash his hands after using the toilet and then handles food that you eat, that's exactly what can happen. Babies' diapers are a good place to find *E. coli*, too (as well as other nasty things that we'll discuss in chapter 5). In 1997, the CDC did an epidemiological study to figure out where you were most likely to get *E. coli* O157:H7 infection, and found that most infections could be traced to living on or visiting a farm, eating under-cooked hamburger meat, or obtaining beef from private slaughtering sources rather than in a conventional store. Contrary to what you might believe from what you read in the papers and hear on TV, eating in a table-service restaurant is a greater infection risk than eating in a restaurant that is part of a major fast-food chain. Maybe this is because the fancier the restaurant, the more they touch your food with their hands, but as far as we know no one has proven this to be the case.

Between 1993 and 1997, 3,260 cases of *E. coli* infection were reported to the CDC, and 8 of them were fatal. As with many infectious diseases, the very young, the very old, and the immunocompromised are most at risk. Normally, healthy people don't die from this infection, and the only treatment for most uncomplicated cases is supportive—that is, you go on

a bland diet and drink plenty of fluids. Antibiotics have little effect, either on the illness itself or in limiting the serious complications. If HUS or TTP is involved, the treatment is quite complicated and includes kidney dialysis, plasmapheresis (withdrawing blood, separating the plasma, and then returning the blood cells to the patient), and other therapies.

There is another interesting characteristic of *E. coli* O157:H7: it is more common in some geographic areas than in others. O157:H7 infections occur in the United States as a whole at a rate of about 2.9 per 100,000 population per year. But in Georgia the rate is 0.6 per 100,000 and in Minnesota it is 5.5—almost 10 times as common. No one is certain, but differences from region to region in the number of direct farm exposures and the consumption of locally processed beef may contribute to this variation. When it comes to *E. coli* exposure, cities are probably safer than rural areas. There is considerable seasonal variation in the occurrence of infection as well—most infections occur during the summer.

O157:H7 is not the only serotype that produces toxin and causes serious illness. In Helena, Montana, in 1994, someone came down with an infection of a serotype called O104:H21, which had all the familiar symptoms of its more famous cousin. Eventually, ten other residents of or visitors to Helena were found to have the same germ. In this outbreak, most of the people affected had consumed milk and milk products from a certain dairy. When they tested the pipes and other equipment used to transport pasteurized milk at the dairy's plant, they found high levels of fecal coliforms, but none of *E. coli* O104:H21, so the source of the outbreak remains uncertain.

Apple cider is often locally produced in small mills and sold unpasteurized. Many people believe it tastes best this way. But there can be problems with it. In 1996, there were three outbreaks of O157:H7, each associated with a different brand of unpasteurized apple cider. In the largest outbreak, 66 people got sick and one died. Using "drop apples"—that is, apples that have fallen to the ground—is common in cider making. But such apples can become contaminated with manure from infected cattle, which may be what happened in this case.

Anyone who gets bloody diarrhea should have a stool specimen cultured—it's the only way to know if O157:H7 is the culprit. Although aggressive treatment is required for HUS and TTP, there's very little you can do about an uncomplicated infection from O157:H7. Antibiotics are not only ineffective, but can actually lead to kidney complica-

tions. Even antidiarrheal agents such as Imodium, Pepto-Bismol, or Kaopectate should be avoided, because they make things worse. Most people simply recover within a week or 10 days. That leaves prevention as the only alternative.

Cooking food well almost always eliminates the threat of infection from *E. coli*—it is rendered harmless at temperatures above about 160°F. If you want to be absolutely sure your food got that hot, you have to test it with a meat thermometer. Pasteurization also makes the germ harmless, so if you never eat pink beef and drink only pasteurized beverages, you can be pretty certain *E. coli* won't harm you. But if you, or those who prepare your food, don't wash their hands before leaving the bathroom, you're taking your chances. Ground beef can still be contaminated after it's cooked, and milk products can be contaminated after pasteurization. This means you shouldn't put cooked hamburgers on the same unwashed plate that held them when they were raw. Anything that touched raw meat—counters, utensils, a meat thermometer, plates, your hands—should be washed with hot soapy water before they touch anything else. You should also keep refrigerated meat separate from other foods.

The Most Common Bacterial Causes of Bad Bellyaches

Since 1989, when it surpassed the number of outbreaks caused by salmonella, *Campylobacter* has been the most common cause of bacterial gastroenteritis in the United States. An estimated 2.4 million people get infected with it every year. It's a gram-negative rod-shaped bacterium, and there are three species of it that cause human disease. *C. coli* causes diarrhea, and *C. jejuni* is one cause of meningitis in infants. The third species, which used to be called *C. pylori* and has recently been renamed *Helicobacter pylori*, is a nasty germ that causes more than 90 percent of cases of peptic ulcer. *C. jejuni* and *C. coli* can turn up in the kitchen. It is less certain where *Helicobacter* comes from. Undercooked poultry has been identified as a major source of outbreaks, and water can be contaminated with it as well.

Why is *Campylobacter* infection suddenly so common? First, it really isn't sudden. In 1886, the bacteriologist Theodore Escherich (after whom *Escherichia coli* is named) first saw something resembling *Campylobacter* in the stool of children who had diarrhea, and various scientists

over the years noticed the organism in other human patients with diarrhea and in animal tissue as well. By the early 1970s sophisticated techniques for culturing the organism had been developed, and it then became quite clear that *Campylobacter* was a very common cause of human illness. So the infection has been around for a long time.

But today the bacterium is flourishing more than ever. Part of the reason is its rapidly developing resistance to antibiotics. The rate of resistant *Campylobacter* is highest in the industrialized world, where antibiotic use is most widespread. Poultry is a significant carrier of the germ, but within two years of the approval of fluoroquinolone to kill the germ in chickens in the United States in 1995, a Minnesota study showed that resistant strains of *C. jejuni* doubled the number of cases of illness.

Although *Campylobacter* can be carried in various ways, including in water, the biggest risk is in handling and eating poultry, particularly eating undercooked chicken. Chickens are chronically infected with it, and almost all raw chicken and other poultry you buy is contaminated with it. Unpasteurized milk is another source of outbreaks, often involving outbreaks after school field trips to farms during the spring. Sporadic infections usually are most prominent during the summer. For poorly understood reasons, there is an unusual age distribution in the infection: infants have the highest rates, followed by young adults. The lowest rates are among young teenagers and middle-aged and older adults. It may be that infants are more susceptible on first exposure, or that they get more immediate medical care and therefore the disease is diagnosed properly in infants more often than in other age groups. Perhaps young adults are more susceptible because their food is more commonly prepared with poor food handling practices. No one really knows.

Infection with *Campylobacter* is usually mild. In cases that are serious enough to get reported to the CDC, which is a very small proportion of all cases, the most common symptoms are diarrhea, fever, and abdominal cramps. Bloody diarrhea is a symptom in about half the reported cases, and, rarely, the organism can cause bacteremia (bacterial infection of the blood) and a bone infection that leads to a form of arthritis called Reiter syndrome, in which pain in the joints can last for months and in some rare cases becomes chronic. About 1 in 1,000 cases of *Campylobacter* infection is followed by a severe nerve illness called Guillain-Barré syndrome, which can cause paralysis. About 40 percent

of Guillain-Barré patients have evidence of a recent case of campy-lobacteriosis. Both Guillain-Barré syndrome and Reiter syndrome are believed to be caused by an autoimmune response to the infection.

The treatment for *Campylobacter* infection is usually just support-ive—you make the patient comfortable and have him drink plenty of fluids with electrolytes to replace those lost in the diarrhea. In very severe cases of dehydration, intravenous methods may have to be used, but this is uncommon. *Campylobacter* infections can be treated with antibiotics—tetracycline, chloramphenicol, and erythromycin alone or in combination are effective—but this usually isn't necessary except in cases where there is very high fever, bloody diarrhea, or more than eight stools in 24 hours. Antibiotic treatment can reduce the duration of illness, but only if it is given quite soon after infection.

To figure out if *Campylobacter* is the source of your problem, you need to have your blood or other bodily fluid tested. The germ can also be cultured from stool, but that requires a more complicated technique. It is especially important to find out if *Campylobacter* is the cause in cases of peptic ulcer. If it is, antibiotic treatment is essential.

As few as a couple of hundred *Campylobacter* bacteria can cause dis-ease—and that amount can easily be present in one tiny drop of raw chicken juice. But it is actually a fairly delicate organism, easily destroyed. It grows best at a temperature between 98.6° and 107.6°F—just about right for the range of temperature inside a human being (or a chicken). If you heat it to more than 131°F for more than a minute or two, it dies. Freezing kills it, too, within a week or so. Moreover, it requires a low-oxygen atmosphere to grow.

Since it is present in stool, *Campylobacter* can be transmitted from per-son to person. This won't happen if everyone washes his or her hands after using the toilet. If you treat all raw meat as if it were infected, you can pre-vent transmission through food as well. You probably can't prevent it, however, by buying "clean" chicken, because there is, for all practical pur-poses, no such thing. The U.S. Department of Agriculture in 1994 and 1995 found *Campylobacter* in more than 88 percent of broiler chickens nationwide, so the last chicken you cooked was almost certainly contam-inated. But you don't get sick 9 out of every 10 times you eat a broiler chicken. In fact, you almost never get sick from eating one. This is partly because you're lucky in that the germ usually doesn't cause disease even when it is present. Of course, it also helps to follow good food prepara-tion practices. This means wrapping raw meats so that juice doesn't drip

out of the package, refrigerating foods promptly, keeping cutting boards clean, washing your hands often before and during cooking chores, making sure you cook everything thoroughly, and avoiding eating raw or unpasteurized foods—the same kinds of precautions you would take to prevent other kinds of food-borne illness.

The Bacteriologist's Namesake

Bacteriologists are quite proud when their name becomes attached to a disease-causing organism. This is not true of other people—American Legion members are not at all happy about having Legionnaire's disease named after them, and homeowners and real estate agents in Lyme, Connecticut, could have done without having their city's name permanently attached to Lyme disease. But Kiyoshi Shiga, a Japanese bacteriologist, must have been extremely pleased to be immortalized in the name of the family of bacteria he discovered in 1870: *Shigella*. That the bacterium causes bloody diarrhea, fever, stomach cramps, and, in one of its forms, death from dysentery in no way diminishes the honor.

There are several different species of *Shigella*. The most common in the United States, causing about two-thirds of shigellosis, is *Shigella sonnei*, or Group D *Shigella*. *Shigella flexneri*, or Group B *Shigella*, accounts for the rest. Other types are rare in this country—including one unpleasant type called *Shigella dysenteriae* type 1, which causes deadly epidemics in many parts of the developing world. Although the illness *Shigella* causes is often referred to as "bacillary dysentery," its effects in industrialized countries are mild, not at all resembling the severe and deadly dehydration characteristic of dysentery.

Shigella are gram-negative bacteria closely related to *E. coli*. Although there are only about 500 cases a year confirmed by the CDC, the actual number of infections is probably much larger, since most infected people recover without ever learning what organism made them sick. Infection is more common among children, particularly among the urban poor and children institutionalized in custodial care. *Shigella* affects only humans and some nonhuman primates, and is very infectious within families. In cases where one kid comes home with the disease, up to 40 percent of the other kids and 20 percent of the adults in the same household will become infected. Usually adults are infected without symptoms, probably because they have become immune.

Infants who are breast-feeding exclusively do not get infected, and this is particularly important in the developing world, where children are often the most common victims of the deadly *S. dysenteriae* type 1.

Shigellosis is frequently transmitted in food. The bacterium can survive the effects of stomach acid quite nicely, and they invade the cells of the colon, moving from cell to cell, destroying them as they go. They do this by producing a poison, shiga toxin, which causes ulcerous sores on the lining of the colon. Mild diarrhea is the most common symptom; severe or bloody diarrhea is rare with the types of *Shigella* common in the United States.

How do you get it? You can round up the usual suspects—people who don't wash their hands after using the toilet—because the germ is present in stool. Vegetables can get contaminated by manure in the fields where they are harvested, flies can breed in infected feces and then contaminate food, and the bacteria can be transmitted by drinking or swimming in contaminated water. Kids who aren't toilet trained are a good source, too, and I'll have more to say about them in chapter 5.

Antibiotic treatment of shigellosis is problematic, and in most cases unnecessary. It's hard to pick an antibiotic because by the time you know for sure by a lab test that you are infected with *shigella*, you're already getting better. Antibiotics at that point wouldn't help at all. If they're given early enough, antibiotics can shorten the illness and shorten the time you are infectious to others, but the germ is already largely resistant to ampicillin, tetracycline, and chloramphenicol. It still responds to a combination of trimethroprim and sulfamethoxazole and to norfloxacin, ciprofloxacin, and furazolidone, but no one wants to use these antibiotics unless absolutely necessary for fear of creating more resistant strains. In most cases, the best treatment is just to try to get comfortable, drink fluids, and wait until it passes. Using antidiarrheal cures such as Imodium or Lomotil is a terrible idea—they slow the action of the intestines, encourage the growth of the organism in the stool, and prolong the fever.

The Case of the Contaminated Hot Dogs

In the late summer and fall of 1998 *Listeria monocytogenes*, a gram-positive bacterium that lives everywhere in the world except Antarctica, caused about 40 illnesses in ten different U.S. states. The CDC in cooperation

with state health departments in Connecticut, New York, Ohio, and Tennessee determined that all these cases came from a single strain of the bacterium in a production run of contaminated hot dogs produced by the Bil Mar Food company and sold under several different brand names. Six of the victims were newborn babies who had gotten the disease from their mothers. Three elderly people and one fetus died. Another 10-state outbreak in 2000 killed 4 people and led to 3 miscarriages, this time caused by turkey meat contaminated with *L. monocytogenes.*

Salmonellosis is much more common than listeriosis, but listeriosis is more commonly fatal. There are an estimated 1.4 million cases of salmonellosis per year in the United States, and 500 people die from it. There are only about 2,500 cases of listeriosis per year, but of those about 500 die. Almost all *listeria* infections come from food. The disease can be transmitted by an infected pregnant mother to her fetus, but there is no other direct person-to-person transmission. As with many food-borne diseases, the victims are most often under 1 year old or over 55 or have immune systems compromised by disease or medical treatment. Healthy adults and children may get infected, but it almost never results in serious illness when they do; they can be infected and have no symptoms at all. If that contaminated turkey and hot dog meat had been heated up properly, the people who ate it probably would not have gotten sick because *L. monocytogenes* cannot survive cooking or pasteurization.

There are six different species of the genus *Listeria*, but only *monocytogenes* causes disease in humans. It's all over the place—not just in food but also in soil, vegetation, marine sediments, and water—and animals can carry it, too, in their intestines, just as humans can, without any symptoms of disease. It has been isolated not only in meat and cheese, but also in raw fish, cooked crabs, raw and cooked shrimp, and smoked fish. Although it will rarely survive cooking above about 140°F for longer than five minutes, it's still pretty durable. It can grow in temperatures ranging from 31°F (i.e., colder than your refrigerator) to 113°F, and it will survive a 10 percent salt solution. Even with its ubiquity and durability, it never bothers most of us. But if you are in one of the high-risk groups, you should take precautions to avoid getting sick from it.

Pregnant women are more likely than others to get listeriosis, but no more likely to suffer serious illness than anyone else. Usually it's very mild, and those who contract it get better. But the illness can cause mis-

carriages, and babies can be born infected, having caught the disease through the placenta. In newborns, the disease can be fatal. Therefore, in addition to taking the usual precautions to prevent any food-borne illness, pregnant women and others at high risk should avoid soft cheeses such as feta, Camembert, and Brie—outbreaks of listeriosis have been traced to such foods. Hard cheeses, cream cheese, cottage cheese, and yogurt needn't be avoided. Although the risk is not high, pregnant women and others might want to avoid deli meats as well, or at least reheat them thoroughly before eating them. If you get a fever or a stiff neck while you're pregnant, you should see a doctor because the only way to tell if it's listeriosis is to test your blood or spinal fluid. If you take antibiotics early enough, infection of the baby can be prevented, although this doesn't always work. Antibiotics are generally not used in listeriosis cases except when the victim is in one of the high-risk groups—an infant, a pregnant woman, or a person whose immune system is not functioning properly. Others recover without medical intervention.

Prevention is the same with *Listeria* as with other food-borne germs: cleanliness is next to healthiness. Washing raw vegetables, cooking meat thoroughly, cleaning off utensils and countertops between uses, and washing your hands frequently and especially after leaving the bathroom will prevent many bad things from happening.

A Germ for All Seasons

If some evil genius wanted to create a really virulent and versatile bacterium, he might model his creation on *Staphylococcus aureus*, a spherical gram-positive organism that grows in clusters that under a microscope look like bunches of grapes. This is a germ that really has what it takes. It can grow in temperatures ranging from about 50°F up to 113°F both with and without oxygen, survive for several days at refrigerator temperatures, and live in a solution that is up to 13 percent salt. It has surface proteins that help it colonize host tissue, and it produces substances called invasins that help spread the bacteria after colonization. Certain biochemical properties of the organism give it high resistance to the phagocytes in the blood that normally ingest and destroy foreign organisms. It releases a toxin called leukocidin that destroys granular leukocytes, one of the types of white blood cells that protects against infection. It produces two other types of toxins called

enterotoxins, one of which causes diarrhea and vomiting and the other toxic shock syndrome. It also puts out a poison that causes blistering and loss of skin in newborn babies, apparently by targeting the specific protein that is involved in maintaining the integrity of the epidermis. While the germ itself is killed by pasteurization, these toxins can still be potent even after they've been cooked at 250°F for up to half an hour. And to top it off, *S. aureus* is highly resistant to antibiotics—some recently discovered strains are even resistant to vancomycin, usually considered the antibiotic of last resort. It's a big problem in hospitals—we'll have more to say about that feature of *Staphylococcus* in chapter 11.

With survival skills like that, you'd think *S. aureus* would be pretty common, and you would be right. In fact, you probably have some right now on your skin or in your nasal passages, throat, or hair. About 50 percent of people do. Animals have it, too. You can live quite healthily with it, provided it doesn't get into the wrong place in your body. It can enter the body through your nose, through wounds, or through IV drug needles. If it does, it can cause a remarkable variety of illnesses: sties in your eyes; boils, carbuncles, impetigo, and other rashes on your skin; endocarditis (an inflammation of the lining of the heart); pneumonia; toxic shock (the same as the toxic shock syndrome women can get from sanitary napkins) and bladder infections; and osteomyelitis (an infection of the long bones). It can also cause septicemia, a generalized blood infection that can be rapidly fatal.

And it causes food poisoning, too, with plenty of vomiting and diarrhea. In the United States, pork (particularly baked ham), fish, chicken, potato or meat salads, and cream-filled bakery products have been implicated. The germ itself, as we noted previously, is killed by pasteurization or by heating the food before you eat it, but the toxins the germ produces are not so easy to destroy. This creates some complications for scientists in establishing that *S. aureus* is the actual cause of any given case of food poisoning. The presence of the organism itself isn't enough—you have to actually establish that the germ you found is producing enterotoxins. Conversely, the lab might find only a few organisms that are the remains of a larger population that produced ample toxin to cause illness. And *S. aureus* isn't the only species that can produce toxins: cases in which *S. epidermidis*, *S. intermedius*, and *S. hyicus* produce poison have been found as well.

Staph isn't like other food poisoning agents—that is, it doesn't actually invade the cells of the stomach and small intestine. Instead, the

enterotoxin it produces acts like a neurotoxin, causing symptoms by attacking the vagus nerve endings in the stomach. The toxin also affects the macrophages, the large cells found all over the body that clean the blood by ingesting damaged red blood cells and other tissue. It causes the macrophages to release interleukin-2, an immune response that leads to inflammation and further unpleasant symptoms.

Most food poisoning from staph occurs during the summer, with mayonnaise frequently the vehicle. You get sick pretty quickly after consuming food contaminated with staph—within two to six hours, nausea, vomiting, and stomach cramps start. Exactly what quantity of toxin you have to ingest to get these symptoms isn't really known, and there are lucky people who ingest contaminated food without getting sick at all. *S. aureus*, like most germs, causes more severe illness in those who are immunocompromised or have some other disease. An average of about 1,400 cases are reported to the CDC every year, and the number of actual cases is much larger. Death from food-borne *S. aureus* is very rare. The CDC reports only one death in the five years from 1993 to 1997.

An Unavoidable Microbe

Bacillus cereus is a rod-shaped aerobic organism that comes in two different forms, neither one of which is fun to have. The one that causes diarrhea doesn't cause vomiting, but the one that causes vomiting also causes diarrhea. The only good news is that either way, you get sick and get well fast: symptoms can start as soon as a half-hour after eating contaminated food, and they only last about 24 hours. Beef, turkey, Mexican food, rice, potatoes, pasta, and shellfish have all been found to contain *B. cereus*, and anyone can get it. It's widely distributed in soil, dust, and air, it's carried by both humans and animals, and it's impossible to avoid. Since small quantities of the germ are harmless, the secret is to keep it from growing into larger quantities that can produce enough toxin to make you ill. You can't really keep it out of food completely in the normal environment of a kitchen, but you can minimize its effects by thoroughly cooking food and refrigerating leftovers, especially cooked rice and other cereals. Like *S. aureus*, *B. cereus* produces a poison that survives higher temperatures than the organism itself—the toxin can still be potent after it has been boiled.

Party Time

On November 8, 1999, a company in Anchorage, Alaska, threw a party for their 500 employees—a nice catered lunch. On November 10, employees started calling in sick—not just a few, but hundreds of them. On the same day, the company notified the Alaska Division of Public Health that they had some kind of epidemic going on, and the health department immediately undertook an investigation. The symptoms the sick people were having—nausea, diarrhea, and vomiting—can, as you now know, be caused by various bacteria that find their way into food. There were at least 200 people who were clearly suffering from some sort of gastroenteritis, and all of them had attended the luncheon. They'd all gotten sick within 96 hours of eating; the average time it took to show symptoms was 33 hours. The average length of the illness was 24 hours—typical for many different bacterial infections. The health department officials were pretty sure where the germ came from: most of the people who got sick had eaten the potato salad. The potato salad had been prepared two days earlier by two employees, one of whom was himself ill. The sick guy had used his bare hands to mix the ingredients in a 12-gallon plastic container. So health officials knew where the bacteria were; now all they had to do was find out what it was. The health department tested about a dozen stool samples from sick luncheon attendees, but they were unable to find any bacteria at all that could have caused illness. Puzzled, they sent 17 specimens, including one from the ill food handler, to the CDC in Atlanta to see if they could figure it out. The CDC performed a complex test called reverse transcriptase-polymerase chain reaction (RT-PCR). The Alaska health department was right—there were no pathogenic bacteria to be found. But in 13 of the 17 samples, including the sample from the food handler, the RT-PCR revealed the presence of the RNA of Norwalk-like virus (NLV). This was an epidemic outbreak of viral gastroenteritis.

Norwalk-like virus refers to a family of viruses closely related to each other. There are several groups distinguished by the kind of immune response they provoke, and they take their names from the places where the first outbreaks occurred. In the United States, there are the Norwalk and Montgomery County groups and the Hawaii and Snow Mountain groups. England has the Taunton, Moorcroft, Barnett, and Amulree agents, and Japan has viruses named for Sapporo and Otofuke. They are all referred to as Norwalk-like viruses.

Only the common cold is reported more often as a cause of viral disease than viral gastroenteritis, and NLVs are the most common cause of viral gastroenteritis—they're involved in more than 40 percent of cases of what is commonly known as "stomach flu." NLVs are ubiquitous and very hardy. They don't even need to be inside a person to survive, but can flourish on environmental surfaces. They're also highly resistant to disinfection and chlorination, which makes them even more insidious.

How do NLVs get into your kitchen? They're probably there right now, but to be infected with them, you have to ingest them in sufficient numbers—they very rarely pass from person to person directly.* The usual method of transmission is what researchers delicately call the fecal-oral route—in fact, the only source of this virus is the feces of an infected person. This kind of unsanitary event can happen in several ways, but usually it's simply because an infected person used the toilet, failed to wash his hands, and then touched food that he was preparing.

Sometimes the virus is transmitted in sewage-contaminated raw shellfish. That's what probably happened to a group of raw oyster lovers in Florida in November 1993. Not everyone who ate the oysters got sick, but of those who did—45 of 71 consumers—10 of them got sick enough to be hospitalized briefly. Everyone recovered, although a few were sick for as long as a week. The investigation by the Florida Department of Health revealed that the more oysters people ate, the sicker they got. This makes sense, because it is known that you can ingest a certain amount of the virus without getting ill—you have to eat enough of it to overwhelm your natural immune defenses. In this case, the health department confirmed that all the food handling was sanitary and that everything was refrigerated properly. All of the oysters had been gathered from waters monitored monthly for fecal coliform, but some of them had been sold at retail by stores not licensed to do so. Florida's health department officials closed the suspect oyster bed, but they opened it again after a week, and no further cases of gastroenteritis were reported.

* "Very rarely," of course, does not mean never. In June 2001, two summer camps in Wisconsin suffered outbreaks of Norwalk-like virus. An investigation revealed no contamination of food or water supplies. Both outbreaks were spread within each camp by person-to-person transmission, probably aided by the close contact of campers in the rustic settings of the camps.

Shellfish are the only foods that come already contaminated with NLVs, but you can easily contaminate anything that is eaten raw. In the 1980s, there were 10 well-documented cases of outbreaks that involved not only clams, but also fruits, salads, eggs, and baked goods. You can gain immunity to these viruses, but it's not permanent (it lasts about two years), and you're only immune to the particular strain that you've been infected with. Unfortunately, there are lots of different strains. Adults and older children are more susceptible than the very young. NLVs almost never kill anyone, but there was one outbreak (it was the Snow Mountain variety mentioned previously) in a retirement community in California in 1988 that did kill two people. This outbreak was probably caused by contaminated food handlers.

NLVs don't multiply in food the way many bacteria do—they only multiply when they get inside you. And they don't survive cooking. The advice may be getting a little monotonous at this point, but if you cook everything thoroughly and wash your hands often, you're a lot less likely to get a bellyache from an NLV.

Not a Plant, Not an Animal—Just Deadly

Seafood is good for you, and many health-conscious people make it an important part of their diet. But some seafood is really bad for you—and we're not just talking about contaminated shellfish. There are certain kinds of food poisoning you can get from fish that can't be prevented even if you wash your hands until they're chapped and cook your food to a cinder. The problem here is neither a bacterium nor a virus. It's nerve poison produced by tiny marine creatures called dinoflagellates.

It isn't even clear whether these things are plants or animals. Zoologists usually think of them as protozoa, but botanists feel that they are algae. In any case, they are usually classified as protists, a kingdom that is defined by what it is not—not animal, not plant, and not fungi; no development from embryos; no cell differentiation. There are more than 2,000 species of dinoflagellates, and they vary considerably in appearance. Most are tiny, but there are some as large as 2 millimeters in diameter. All but about 10 percent are marine plankton. They all have two flagellae, movable strands of protein, which they whip back and forth in order to propel themselves through water. Some species

are responsible for "red tide," the "algae blooms" that occasionally pollute marine waters. Many are bioluminescent—they glow in the dark. They are single-celled, and they reproduce by mitosis, that is, by splitting into two identical organisms. Some dinoflagellates are photosynthetic—they produce food by manufacturing it with the aid of sunlight. Others live symbiotically with other marine species such as corals or anemone. And about half of them consume other plankton, capturing them either by snaring or stinging. The stinging is effective because many species of dinoflagellates produce paralyzing nerve poison. These poison-producing species (and others) serve as food for herbivorous fish that live near the shore, especially in tropical reefs, and the fish accumulate in their flesh a nerve poison called ciguera toxin, or ciguatoxin, without themselves being killed by it.

These herbivorous fish are not usually eaten by humans, but they are eaten by barracuda, grouper, amberjack, surgeonfish, sea bass, and Spanish mackerel, which accumulate the poison in their flesh. Humans do consume these fish. The larger of these fish, like barracuda, are even more likely to be poisonous than the smaller ones, and the poison in all contaminated reef fish is especially concentrated in the head, viscera, and roe. The poison is stealthy and durable. It's odorless, colorless, and tasteless, and it's not destroyed by cooking or freezing. There isn't any lab test that can detect the stuff once it's inside you, so the only way to be sure that ciguatoxin is what's behind your symptoms is by finding the poison in a leftover piece of the fish you've been eating. And here's more bad news: there's no antidote for the poison, and no treatment for the symptoms.

And what symptoms! Three to six hours after you eat the stuff, you get watery diarrhea, nausea, and abdominal pain that in most cases, but not all, lasts about 12 hours. Then there are the neurologic problems: you start feeling pins and needles around your mouth and in your fingers and toes, you get a rash; hot things feel cold, and cold things hot. Alcohol, exercise, and sexual intercourse make the symptoms worse. In some unlucky people, blood pressure drops, breathing slows, and coma ensues. The stomach upset can in some cases last as long as a week, and the neurologic symptoms can go on for months.

Ciguatoxin poisoning isn't that common—there are an average of about 40 cases a year reported to the CDC—and people almost always recover from it without permanent aftereffects. You might be able to avoid it if you knew where your fish came from. The problem, though,

is that it's almost impossible to know which areas are currently contaminated. There are some tropical islands where one reef contains inedible fish, but a nearby reef is perfectly okay. Of course, most of the time it's impossible to know where your fish came from anyway—fish is shipped all over the world, and you usually don't even know if they're from tropical waters or not, much less whether they came from a contaminated reef. The solution may be simply to avoid eating the species mentioned above that can be contaminated by dinoflagellates. Pellagic species, the fish that swim in the open ocean like tuna, dolphin fish, or wahoo, have never been found to carry ciguatoxin, nor have freshwater fish.

There's no ciguatoxin in tuna, but tuna can still make you sick. Scombroid poisoning, a bacterial illness, is yet another disease you can get from seafood. Scombroid isn't the name of a bacterium, but the name of the group of mackerel-like fish that can be contaminated with various histamine-producing bacteria (*skombros* is Greek for mackerel). These include mackerel, tuna, swordfish, bonito, and albacore. But the bacteria has also been found in nonscombroid species such as dolphin fish, amberjack, and bluefish.

These histamine-producing bacteria grow in warm temperatures, and the infection of the fish can begin after they are caught. This is particularly true in the fishing method called long-line fishing. This consists of suspending a fish line that can be up to 60 miles long and have 3,000 or more hooks on it. Fish are hooked on these lines and can stay in the water for up to 20 hours while the lines are collected and the fish landed. During this period, they can be colonized with bacteria. The formation of the toxin can also occur at any point after the fish is landed and before it is eaten. It's easy to prevent the toxin from forming: land the fish quickly and refrigerate it at 32°F until it's cooked. But not everyone who handles your fish from the time it's caught until the time you eat it does this, and the CDC gets reports of about 80 people a year who get sick from scombroid poisoning. There are probably many more unreported cases as well.

The symptoms start very quickly after you eat contaminated fish: a burning sensation in the mouth, a metallic taste, facial flushing, body rash, nausea, diarrhea, and headache can start as fast as five minutes after you swallow the fish. The symptoms usually resolve by themselves after a few hours, and although medical treatment is usually not necessary, epinephrine, cimetidine, or diphenhydramine can help minimize symptoms.

This histamine is the same stuff that causes hay fever, but in fish it's much more potent. This is probably because there are various potentiators of histamine present in fish that's starting to go bad, including putrefactive amines such as the alarmingly named putrescine and cadaverine. These substances seem to inhibit the ability of the intestines to metabolize histamine, which makes the histamine much more effective in causing symptoms.

At the Raw Bar

Never eat oysters in a month that does not contain the letter R. Or so they say, and they may be right, at least partly. In the summer of 1997, a lot of people ignored this advice, and a lot of people got sick.

It began in July in British Columbia, where the British Columbia Center for Disease Control (BCCDC) got some samples from nine patients who'd come down with gastroenteritis—the usual nasty combination of diarrhea, abdominal cramps, nausea, vomiting, fever, and bloody diarrhea. All the samples contained a gram-negative bacterium called *Vibrio parahaemolyticus*. They wanted to learn where this was coming from, so they tracked down eight of the nine people who'd given them samples to find out what they'd been eating. Seven of the eight had eaten raw oysters. By the end of July, the BCCDC issued a warning telling people not to eat raw or undercooked oysters. By mid-August, they'd figured out where the contaminated oysters were coming from, and the Federal Department of Fisheries and Oceans closed all British Columbia coastal waters to oyster harvesting. They kept them closed until September 12, and no further cases of vibrio infection were reported.

In the meantime, the Washington Department of Health (WDOH) was getting similar reports. About two weeks after the British Columbia infections were reported, the WDOH advised commercial harvesters to refrigerate oysters within four hours after gathering them, and a week later issued a public advisory to cook all oysters thoroughly before eating them. When the WDOH interviewed people who were infected, they found that 90 percent of them had eaten raw oysters. The WDOH closed Washington oyster beds on August 28, then reopened them on September 15, after which no additional cases of infection were reported.

Oregon and California were next. The CDC started finding cases in these states, and issued similar warnings and oyster bed closing orders there. Oregon closed their infected oyster beds on August 26, and no further cases were reported. But California didn't bother to close their beds, which was too bad for the citizens of California, because infections there continued through December. Eighty-three Californians got sick. By the time the whole thing was over, 209 confirmed cases of *V. parahaemolyticus* infection had been reported in the Northwest. Two people had to be hospitalized, and one died.

V. parahaemolyticus loves warm water, and it is possible that the increasing number of outbreaks of disease—there have been four multistate outbreaks since 1997—is one more unpleasant result of global warming. The surface temperature of the Pacific coastal waters between May and September 1997 ranged from 54°F to 66°F—two to nine degrees warmer than during the same period in 1996. The same observation was made in an outbreak in summer 1998 in Long Island Sound: the mean temperature of the water that summer was 77.2°F, compared to 74.1° in 1997 and 69.4° in 1996. These epidemics also called into question the level of *V. parahaemolyticus* that is safe to consume. It is known that the more you eat, the sicker you get, and that small amounts are harmless. But the level of contamination that caused illness in these outbreaks was less than 200 cfu per gram, when the level at which closing oyster beds is recommended is 10,000 cfu per gram. (CFU stands for colony forming unit, or a single cell capable of reproducing to form a colony.) The CDC has affirmed that this standard must be revised.

Other species of *vibrio* have also been implicated in gastroenteritis outbreaks. *V. vulnificus*, for example, causes most of the food-related deaths in Florida (almost all of these people have some underlying illness, usually liver disease, or are immunocompromised). These infections also come almost exclusively from raw oysters, and typically in the summer months.

Vibrio infections don't come from eating oysters that grew in "dirty" water. Vibrio is a naturally occurring bacterium in seawater, not a result of pollution. Raw oysters are delicious. Eating them holds a small chance of making you sick. If you've got liver disease or are otherwise ill or immunocompromised, it's probably a good idea just to stay away from raw oysters completely. But the risk that a healthy person will get sick from eating raw oysters is very small indeed, and even if

you do get sick, the illness is likely to be mild and pass quickly. Many healthy oyster lovers will probably be willing to take their chances.

Even Pasteurization Couldn't Kill It

The CDC couldn't figure out why, but in June and July 1982, almost 900 people in Mississippi, Tennessee, and Arkansas came down with diarrhea, abdominal pain, and in most cases fever from an infection of *Yersinia enterocolitica*. The best guess was that it came from a single milk plant in Memphis that distributed a large amount of pasteurized but nevertheless contaminated milk. The problem with yersinia is that if it is present in large enough numbers, some of it can survive pasteurization. Then it grows well at refrigerator temperatures. This is a bad combination of characteristics in a foodborne pathogen.

By 1990 infections caused by *Y. enterocolitica* emerged from a widely distributed reservoir (that is, a place where a germ lives more or less permanently) in swine. Chitterlings—pig intestine, which is often served as a holiday dish by southern Black families—were a source of a large outbreak in Atlanta around Christmas that year. The potential for transmission from food handlers to children is the strongest, so people who prepare chitterlings (this is a labor-intensive and time-consuming process) should not take care of kids while the food is being made.

Y. enterocolitica is a rod-shaped bacterium in the same family as *Y. pestis*, the germ that causes Black Death. While the main source of the germ is pigs, it is also found in rodents, rabbits, sheep, cattle, horses, dogs, and cats. Children are more susceptible than adults, and the infection is more common in the winter. Fortunately, infection with *Y. enterocolitica* is not very common: only about 1 in 100,000 people get it each year. Still, if you're that one or your kid is, it can be very unpleasant. Most cases simply resolve by themselves without any intervention after a few days of stomach upset. In more severe cases, antibiotics can be used. Very rarely, some people get pains in the knees, ankles, or wrists that can last up to six months. Some, especially women, can get a rash on the legs and trunk, but it usually goes away in a few weeks. *Y. enterocolitica* can be avoided if you don't eat undercooked pork, if you wash your hands after contact with animals or raw meat, and if you are very careful about washing before touching kids or their toys, bottles, or pacifiers.

A New Germ for the Twenty-First Century

A new infectious agent, a protozoan parasite called *Cyclospora cayeta-nensis*, was first identified and named in 1979. It was previously thought to be a kind of cyanobacterium. By 1993, it had been found in the United States. The first U.S. report of a diarrheal outbreak caused by the organism didn't appear until 1995. Then in the first six months of 1997, 21 outbreaks were reported to the CDC. Raspberries imported from Guatemala were the source of the infection. Another outbreak occurred in Ontario in 1998, also attributed to Guatemalan raspberries. Mesclun lettuce and fresh basil (that were not from Guatemala) have been implicated in other outbreaks, and many sporadic individual cases have been reported. Contaminated water is another way the organism can be transmitted. The increase in reported cases is at least partly attributable to better methods of detecting the organism in stool samples, but there is probably a real increase as well.

Cyclospora is neither a bacterium nor a virus—it's a unicellular parasite. It isn't immediately infectious after it's excreted. This is because the organism is excreted in the form of oocysts, which have to sporulate—that is, "hatch," in a sense—before they are capable of infecting anyone. This requires several days or even weeks to happen, and it requires the right conditions. But if all goes according to plan (the parasite's plan, not yours), you ingest the spores with your raspberries, lettuce, or fresh basil. Although it can sometimes cause loss of appetite and weight loss, it will never be mistaken for a diet aid: along with the weight loss you get very watery, often explosive diarrhea and sometimes bloating, stomach cramps, increased gas, vomiting, muscle aches, and fatigue. It takes about a week after you are infected for the symptoms to begin. The diarrhea can go away for a while and then come back. If it isn't treated, the illness can last from a few days to as much as a month or longer, with multiple relapses. The treatment is a combination of two antibiotics, trimethroprim and sulfamethoxazole, usually sold under the brand names Bactrim, Septra, or Cotrim. These are sulfa drugs, and if you're allergic to them, you'll have to do without drug treatment—there are no substitutes. Rest and fluid intake are also required.

Cyclosporiasis doesn't pass directly from one person to another (from an infected person who didn't wash his hands after using the toilet, for example) because the parasite needs time to sporulate before it

becomes infectious. For *Cyclospora cayetanensis*, humans are just one step in a life cycle—we ingest sporulated oocysts, the organism reproduces inside us by asexual reproduction in our gastrointestinal tracts, we excrete unsporulated oocysts, the oocysts sporulate in the environment. Then another human (or the same human, for that matter) ingests them. No one knows if this process goes on in animals as well. Cyclosporiasis isn't one of the more common causes of food poisoning—in 2000, the CDC estimates, only about one in a million people came down with it.

Minimizing Kitchen Risk

So your kitchen is full of germs. What are you supposed to do about it? Most people probably do little except keep the place clean—they wipe up surfaces so that things look neat, put leftovers in the refrigerator, and get the garbage out of the house so it doesn't start to stink. In a survey undertaken in 1996 and 1997 in five different states—California, Connecticut, Georgia, Minnesota, and Oregon—1.5 percent of respondents drank unpasteurized milk, 1.9 percent ate raw shellfish, 18 percent liked their eggs runny, 30 percent preferred pink hamburger, and 7 percent of people confessed to not washing off their cutting boards after cutting raw chicken. Most of these people don't get food poisoning from anything they eat in their kitchens, but every once in a while they do. The Food and Drug Administration has some ideas about how you can minimize the risks of catching something unpleasant in your kitchen.

First, check the temperature of your refrigerator. Most people put the thermostat on some middling number and forget it. But if your refrigerator is warmer than 41°F, you're giving germs a place to grow. In most refrigerators, the thermostat setting doesn't tell you what the temperature is. Put a thermometer in there and find out. Then reset the thermostat if you have to. Try it on several different shelves—the temperature can vary from place to place inside the refrigerator.

Put leftovers away immediately. Don't leave food out unrefrigerated for more than two hours—and if it accidentally happens that you do leave it out this long, throw it out. After three to five days in the refrigerator, most leftover food has about had it. Don't eat it after that.

Here's something that probably very few people do, but the FDA recommends: sanitize your sink drain, disposal, and connecting pipes.

This isn't as hard as you think—all you have to do is periodically, say once a week or so, pour a quart of water with a teaspoon of chlorine bleach in it down the drain.

For most people, there's no temptation to crack open an egg and eat it raw. But you might be tempted to taste your homemade cookie dough before you make cookies out of it. Don't—there's raw egg in there, and eggs have to be cooked to be sure they're free of *Salmonella enteriditis.* Homemade eggnog, mayonnaise, ice cream, and cake batter may all contain raw eggs. Commercial preparations of any of these things are safer—they're made with pasteurized eggs.

Wiping up with plain water won't kill any germs. In fact, it's a good way to spread them around, because a wet sponge is a good place for bacteria to breed. But using bleach, or even just soap and hot water, can go a long way to reducing bacteria (though it won't eliminate them). Commercial disinfectants (see chapter 12) may work, but they may not, and a diluted solution bleach and water is just as or more effective and almost always cheaper.

Defrost food in the refrigerator—it takes longer, but it's less likely to allow the growth of bacteria. Defrosting in the microwave is okay, too, as long as you cook it right after you defrost it. Leaving stuff on the counter to defrost is asking for microbiological trouble. Same with marinating food—do it in the refrigerator, not at room temperature.

And of course: Wash your hands often—probably the single most effective infection control technique there is, in the kitchen and everywhere else.

3

Toilet Training

WASHING HANDS IS
THE BEST REVENGE

John Lynn, of Austin, Texas, thinks people don't wash their hands often enough, and he thinks it's about time someone did something about it. According to Patent Number 6,147,607, Mr. Lynn has designed a way of preventing people from slipping away from the bathroom without washing. He proposes a mechanism attached to a toilet handle that would ooze out some harmless but unpleasant stuff onto the hand of anyone who used it. This substance—paint, dye, or grease of some sort—would leave a stain that would have to be washed off with soap and water. In a hospital, the substance could be one that could only be cleaned off with antibacterial soap or alcohol. Although he doesn't say so in the patent application, presumably Mr. Lynn would make the substance red, giving profound new meaning to the expression "caught red-handed." He does say that in a restaurant, where employers might want to monitor their employees but not their customers, the substance could be one that is invisible except under ultraviolet light.

We can make fun if we want, but John Lynn is right: people don't wash their hands often enough. In September 2000, the American Society for Microbiology reported the results of a survey that found that of 7,836 people observed in five American cities (Chicago, Atlanta, New York, New Orleans, and San Francisco), only about two-thirds washed their hands after using a public toilet. For some reason, Chicagoans

US006147607A

United States Patent [19]

Lynn

[11] Patent Number: **6,147,607**

[45] Date of Patent: *Nov. 14, 2000

[54] **METHOD AND APPARATUS FOR HELPING TO ASSURE THE WASHING OF HANDS**

[76] Inventor: **John M. Lynn**, 8616 Mendocino Dr., Austin, Tex. 78735

[*] Notice: This patent is subject to a terminal disclaimer.

[21] Appl. No.: **09/371,825**

[22] Filed: **Aug. 11, 1999**

Related U.S. Application Data

[63] Continuation-in-part of application No. 09/170,172, Oct. 13, 1998, Pat. No. 6,031,461.

[51] Int. Cl.[7] .. **G08B 23/00**
[52] U.S. Cl. **340/573.1**; 222/39; 106/31.32; 424/9.71; 428/41.1; 422/186; 422/291; 292/347
[58] Field of Search 340/573.1; 222/39; 106/31.32; 424/9.71; 428/61.1; 422/186, 291; 292/347

[56] **References Cited**

U.S. PATENT DOCUMENTS

1,491,780	4/1924	Abbott	292/347
1,783,097	11/1930	Polcari	292/347
2,044,904	6/1936	Hcisig .	
2,527,955	10/1950	Pagel	422/291
3,314,746	4/1967	Millar	422/186
3,877,005	4/1975	Apgar	340/304

4,431,628	2/1984	Gaffar	424/9.71
4,832,942	5/1989	Crace	428/41.1
4,856,140	8/1989	Visco et al.	16/412
5,043,013	8/1991	Kluger et al.	106/31.32
5,810,201	9/1998	Besse et al.	222/39
5,812,059	9/1998	Shaw et al.	340/573.1
5,870,015	2/1999	Hinkel	340/573.1

FOREIGN PATENT DOCUMENTS

567512 1/1933 Germany .

OTHER PUBLICATIONS

International Search Report for PCT/US99/23560, Jan. 21, 2000.

Primary Examiner—Benjamin C. Lee
Attorney, Agent, or Firm—Baker Botts L.L.P.

[57] **ABSTRACT**

A method and apparatus are disclosed for helping to assure the washing of hands. An easily identifiable substance is provided which can be removed by washing, and a marking mechanism is coupled to the easily identifiable substance. A hand of a person is then marked with the easily identifiable substance when the marking mechanism is triggered. In one embodiment, a flush mechanism of a toilet or urinal is equipped with the marking mechanism, and the marking mechanism is triggered when a person flushes the toilet or urinal. In another embodiment, a door handle is equipped with the marking mechanism, and the marking mechanism is triggered when a person uses the door handle.

28 Claims, 1 Drawing Sheet

FIG. 1

Figure 3 The first page of the patent for A Method and Apparatus for Helping to Assure the Washing of Hands. This is a copy of an actual U.S. government document for which we paid $3.00. We do not make these things up. *U.S. Patent Office.*

were more inclined to wash than other people: 83 percent of them washed. New Yorkers were the least interested in hand washing, with only 49 percent of them bothering.

A Nice Warm Bath

Hand washing is unquestionably a good idea, and we'll have more to say about that below, but at the same time, personal hygiene may not be all it's cracked up to be. Bathing and showering, for example, are generally pleasant activities, but as far as infection control is concerned, they probably don't do much good.

It is true that the decline in infectious disease between 1900 and 1940 was accompanied by an increase in showering and bathing, but it's very hard to estimate the contribution of this activity to the general improvement. There are too many other factors: improved waste disposal techniques, chlorination of water, changes in commercial food handling, increased understanding of nutrition, and public education programs all helped considerably. Aside from hand washing, there isn't much scientific evidence to link personal hygiene to disease prevention. While it may seem that a daily bath is a sanitary thing to do, whenever anyone looks at the practice scientifically, they discover that it doesn't do much of anything. A 1979 study reported in the *Archives of Dermatological Research* is typical: daily baths, it found, have no effect whatever on the composition of the normal bacterial flora of the skin.

Even in hospitals, bathing patients with medicated soap before they have operations doesn't help prevent postoperative wound infections. A 1983 study looked at more than 5,000 patients in three different hospitals and found that bathing with antibacterial chlorhexidine detergent was no more effective than bathing with ordinary soap and water in preventing infections of surgical wounds.

On the contrary, there is some evidence that not only do bathing and showering have little or no effect on infectious disease, but they may in fact have some role in increasing it. There are several good studies that show that bathing and showering increase the dispersal of skin bacteria into the environment, and as a result surgical personnel are now discouraged from showering before entering an operating room. The amount of bacteria on the skin varies considerably from person to person, but it is quite consistent on each person over time. Even if you

go without showering or bathing for many days, the amount of bacteria on your skin doesn't change, either in quantity or quality.

Soap and Water

It may be some time before Mr. Lynn's clean-hands identification mechanism is in widespread use, but in the meantime, you might want to wash your hands more often anyway, whether you can be caught at it or not. There is indisputable evidence that washing hands—we're talking about the general public here, not surgical personnel—is very effective in preventing the spread of disease, especially of pathogens that are transiently on your hands (like cold germs, which we'll discuss in more detail in chapter 8) and those that are spread by the fecal-oral route. Plain soap and water is just as effective as any other agent for this purpose—if only people would wash frequently enough and long enough when they do wash. Unfortunately, they don't. To wash your hands well enough to prevent disease transmission you have to use soap and warm water, and you have to rub your hands together vigorously for 10 to 15 seconds, working up a good lather and covering all the surfaces of your hands and fingers. Then you have to dry your hands with a paper towel (preferred) or a clean towel or air dryer. Then you're supposed to use a paper towel to turn off the faucet. And you're supposed to do this before preparing food; before eating; after sneezing, coughing, or blowing your nose; after using the toilet or touching a pet; whenever your hands are obviously dirty; and after cleaning activities. And that's just for your average person—for some people the hand washing has to be more elaborate and more frequent. Food service employees, for example, in addition to the washing we're all supposed to do, are supposed to wash their hands before beginning or returning to work, immediately after preparing food, after touching any bare human body parts other than clean hands and arms, after handling money, and every time they switch between working with raw food and working with ready-to-eat food.

Let's face it: very few people are going to run in and wash their hands every time they cough or sneeze, and even if they do they aren't very likely to follow such an elaborate hand-washing regimen. Using the toilet is another matter, though, and it seems reasonable to expect that people will wash their hands after moving their bowels. People

overwhelmingly say they wash after they use the toilet, but observation proves they often don't. The American Society for Microbiology, in the survey mentioned above, found that while 95 percent of people say they wash their hands after using a public toilet, when you actually watch them in action, only 67 percent do. Men are worse than women—92 percent of them claimed to wash their hands, but only 58 percent did. With women, the gap was smaller, but still surprising: 97 percent said they did, 75 percent actually did. For some reason, people who make less money wash more often than people who make more, or at least they say they do. In every category—before eating, after petting a dog or cat, after coughing or sneezing, after handling money—people who make less than $35,000 a year report more hand washing than people who make more, and people with only a high school diploma report more hand washing than those who went to college.

It is even possible to infect yourself with your own stool. Pinworms are often recirculated in this way by children. If stool gets into your urinary tract, this can cause infection as well. And stool can also infect sores or abscesses around the rectum. People who are immunocompromised can be infected with their own intestinal flora through breaks in the intestinal mucosa.

In any case, germs don't care how much money you make, and they don't care whether you went to college. Everyone has them, and anyone can spread them. The stratum corneum, as the topmost layer of your skin is called, is a mixture of flattened dead cells and lipids that forms a tough protective outer layer. Healthy people shed particles of this layer every day—10 million of them—and about 10 percent of them contain viable bacteria. Men give off more than women, and while there is considerable variation from person to person, bathing routines don't have much effect: a person using exactly the same hygiene measures as another can nevertheless disperse more than five times as much bacteria. You are a walking bacteria factory, spreading germs wherever you go, and there isn't much you can do to prevent it. Nor do you have to for the sake of your health or anyone else's.

If you've read chapter 2, you know that the fecal-oral route is a common one for the transmission of disease, and that if you touch food after using the toilet without washing your hands, you are asking for trouble. Human feces, even the feces of perfectly healthy people, are filled with dozens of different species of disease-causing microorganisms.

The natural disgust at putting feces anywhere near your mouth finds justification in scientific fact: you can get really sick by doing so.

Approximately a third of the weight of human feces is bacterial debris—100 billion bacteria of 75 different kinds are excreted by every healthy person every day. Among these normally excreted bacteria are many pathogens. And if you're not healthy, your feces can be even more dangerous. Improvements in plumbing and waste water disposal in the first 40 years of the twentieth century were a major contributing factor to the sharp reduction in infectious disease simply by keeping feces-polluted water away from people's lips.

Okay, so we know it's important to wash your hands when you leave the bathroom. But is there such a thing as being too clean? Washing your hands too much? Alas, yes.

Every time you wash your hands, you change their bacterial flora. Most of these changes are temporary—after a few hours, your skin has returned to its normally bacteria-laden state. But if you wash your hands repeatedly (health care workers, for example, may wash their hands as often as 10 or 15 times a day), and especially if you use strong detergents, you can cause contact dermatitis—irritated, damaged skin. Damaged skin harbors and disperses more disease-causing bacteria than undamaged skin, which is clearly not the desired result in hand washing. Skin condition under these circumstances takes on great importance.

Antiseptic soaps reduce bacteria on the hands, but not beyond a certain level. And using them instead of soap for washing may actually make things worse. In one study, the number of organisms spread from the hands of nurses who used antibacterial soap actually increased, probably because of skin irritation. In another, nurses with damaged skin did not have higher microbial counts, but did have a greater number of colonizing species and were more likely to be contaminated with drug-resistant staphylococcus than nurses with undamaged skin. So it's probably best to find some moderate position: enough hand washing to encourage infection control, but not so much as to damage the skin.

Antibacterial soaps are not necessary for the general public for routine bathing, although they might be useful in special situations, such as when skin infections are present or when preparing to perform surgery. Even for food handlers, the use of antibacterial preparations has to be weighed against the possibility of encouraging the prolifera-

tion of resistant strains of bacteria. As a substitute, for health care workers and others, alcohol can be used as a disinfectant. Alcohol kills a large variety of germs and doesn't lead to the formation of resistant strains. Alcohol-based hand washing products are widely available over the counter and have their own problems, about which we'll have more to say in chapter 12. If you wash your hands often, whether with antibacterial preparations or not, the use of hand moisturizers can be beneficial for skin health.

Fearless Scientists at Work

So much for hand washing and bathing. Now what about those porcelain facilities? What microbiological horrors lurk in the toilet bowl, the sink, and the bathtub, and what are you supposed to do about them? There are plenty of germs that accumulate in those places, but as far as your health is concerned, your kitchen is almost certainly more dangerous than your bathroom.

Much of the research done on bathroom microbiology is done in hospital or other institutional settings, where infection control is a considerably greater problem than it is in your bathroom at home. Nevertheless, we can draw some conclusions from this work, and from the smaller body of research that is devoted to home toilet facilities. Let's start with one of the more common germs, salmonella.

Salmonella is not considered a contagious disease. You get it in the various ways you read about in chapter 2, mostly from contaminated food, and you don't usually pass it on to members of your family. But a recent British study showed that in a home where someone with diarrhea caused by salmonella used the toilet, there were plenty of salmonella bacteria left in the bathroom—even for as long as four weeks after the diarrhea had stopped. It was a small study—they tested six households—but they found that in four of them the toilet bowl rim and the scaly biofilm adhering to the toilet bowl surface below the water line were both contaminated with *S. enteritidis*. Even using toilet cleaning fluids didn't get rid of them. On dry areas—the toilet seat, the handle, and the door handle—no salmonella were found. Then the researchers set up an experiment to mimic toilet use by a salmonella-infected person with diarrhea. A microbiological researcher was bound to come up with this idea, but it took two researchers at the Pharmaceutical Sciences

Institute, School of Life and Health Sciences in Birmingham, England, to actually follow through with it. Anyway, we leave to your imagination the details of the way the experiment was set up and carried out, but they learned that under experimental conditions, flushing a toilet full of salmonella-infected diarrhea caused salmonella bacteria to contaminate the toilet seat and toilet seat lid. Bacteria were also isolated from an air sample taken immediately after flushing, suggesting that they could land on and contaminate other surfaces in the bathroom. In one case, they found bacteria in the toilet bowl water. The authors conclude that there is "considerable risk of spread of salmonella infection to other family members via the environment." Judging from how rarely salmonella is in fact passed from one family member to another, this seems rather overstated. While it's true that a salmonella-infected person who doesn't wash his hands after he uses the toilet can contaminate stuff he touches, including food that someone else might eat, it must be very rare indeed that anyone contracts salmonella poisoning from the small amount of bacteria left in a bathroom recently used by that person. With some organisms like shigella, if you ingest just one or two of them you can get infected. But salmonella requires a good-sized dose—you have to eat about 100,000 of them to get sick.

To get infected with salmonella from using a toilet, you would probably have to scrape your hands along the edge of the toilet bowl or wipe them on the surface of the bowl beneath the waterline. You would have to do this soon after the infected person had used the toilet. Then you'd have to put your hands—before washing them, of course—in or near your mouth to get the germs inside you. And you'd have to get a large enough dose to cause disease. This seems an unlikely series of events, at least for adults. On the other hand, there is a Chilean study that shows that typhoid fever (which is caused by a bacterium of the same family as the one that causes salmonellosis) was transmitted among the families of kids in a manner very much like this. The researchers had two groups of 40 kids, 20 of whom from each group had had typhoid fever. One group was of high socioeconomic status, one of low. They showed by statistical means that dirty toilet bowl rims and nail-biting were very good predictors of who would get sick. And they also showed that high socioeconomic status was no protection: the rich kids got typhoid fever just as easily as the poor kids, adding more evidence to the assertion earlier in this chapter that germs don't care how much money you have, and suggesting further that they don't care how much money your parents have, either.

These researchers are not the only microbiologists to indulge an interest in toilet bowl cleanliness by conducting scientific studies of what happens when a toilet flushes. Some years ago, a group at the Baylor College of Medicine in Houston tried "seeding" a toilet bowl with a strain of *E. coli* harvested from domestic sewage and with a type of poliovirus to see what would happen when they tried to flush them away. They put a sample of *E. coli* (carefully measured to simulate its concentration in real fecal matter) into a toilet bowl, then began sampling the resultant brew for bacterial concentration. They flushed, measured, flushed, measured, and flushed again and measured again. They found that the first flush eliminated most of the bacteria, but subsequent flushes didn't improve things much—in fact, "there was often an increase in the number of residual organisms detected in the bowl." The same thing happened when they did the flush tests with a sample of poliovirus. They concluded that the germs were sticking to the porcelain surface of the toilet bowl. When they tried eliminating them by using a toilet bowl disinfectant, it worked quite well. But the toilet bowl cleanser worked no better than just scrubbing the bowl with a toilet brush.

Then they decided to see what happens to water droplets when you flush. Here their ingenuity really came into play. First they dyed the toilet bowl water purple. Then they covered the bowl with a sheet of white absorbent paper so that any drops would fly up onto the paper and make purple stains that could be counted. Tank toilets produced 27 to 104 drops per flush, in a random pattern on the paper. Valve toilets only produced 7 to 10, clustered toward the rear of the bowl. So they knew from this that droplets were certainly flying out of the bowl with every flush. Then they covered the toilet seat with some gauze, contaminated the toilet water with *E. coli*, flushed, and collected the droplets on the gauze. They found that the number of bacteria ejected onto their gauze was directly proportional to the amount of bacteria with which they had contaminated the toilet bowl water. Then they put their gauze on other bathroom furniture—the sink, the faucets, the floor, and so on. Each time they found that they could culture out bacteria until about four hours after flushing. After that time, the bacteria had pretty much disappeared. The results were more or less the same when they tried the procedure with the virus, but the virus disappeared more quickly—within about two hours.

There are other studies, too, that show that flushing a toilet can put germs into the air that can then settle on surfaces in the bathroom. But

this is probably not a very serious threat, and carefully sanitizing your bathroom with antibacterial cleansers is unlikely to do much for your health. The CDC has concluded that even in hospitals, where there is considerably more risk than in a bathroom in a home, environmental surfaces contaminated with germs are not associated with transmission of infection to either staff or patients. For that, you need direct hand contact, which is why hand washing is so important—more important than sanitizing the toilet bowl.

This doesn't mean you should stop cleaning your bathroom. Obviously there are esthetic reasons for doing so, and discouraging the presence of germs there is certainly a good, even if not lifesaving, idea. Viruses generally don't survive well on environmental surfaces, but bacteria can and do. Gram-positive bacteria such as staphylococcus survive best in dry conditions, while gram-negative bacteria and fungi prefer moist areas. The bathroom has to be cleaned only with ordinary soap and water—you can use a disinfectant, but the actual scrubbing of surfaces is more important for getting rid of germs than the use of any antibacterial detergent. Washing the floor with disinfectants is probably a waste of time. Disinfection of floors has no advantage over washing them with soap and water, and anyway floors become contaminated as soon as you're done disinfecting them because of airborne microorganisms and even from the mere presence of human bodies and shoes.

The cloth, sponge, or mop that you clean with can itself be a means of moving germs around. If you want to be really careful, you should wash your cleaning cloths or mops out between uses and let them dry before using them again. This minimizes the possibility of having contaminated cleaning equipment. Some people will be willing to make this kind of effort; some won't. There isn't much proof that the former will wind up any healthier than the latter.

One study done at Georgetown University School of Medicine found that even bars of soap in the bathroom are contaminated with bacteria and that wet soap has more bacteria than dry soap. The major bacteria species isolated from the bars of soap were staphylococcus species and enterobacteria. Again, though, the authors were unable to draw any epidemiological conclusions from this—that is, they couldn't prove that these bacteria were actually making anyone sick.

In general, if you're looking for bacteria on environmental surfaces, you're more likely to find them in the kitchen than the bathroom. When they examined kitchens and bathrooms carefully, a group of Ari-

zona researchers found that the *least* contaminated place was the toilet seat, and that moist environments in the kitchen—the sponge, the dishcloth, the kitchen sink drain, and the kitchen faucet handle—were the best homes for bacteria. The most contaminated place in the bathroom was the sink drain, another nice wet place for bacteria to thrive.

And You Put It in Your Mouth?

Shocking but true: your toothbrush is dirtier than your toilet seat. It is hard to predict exactly which germs are on your toothbrush right now, but staphylococci, coliforms, pseudomonads, streptococci, and at least one fungus, candida, have all been cultured from used toothbrushes. In one Australian study, staphylococci and streptococci were the most commonly found pathogens on used toothbrushes. Even if you have a brand-new toothbrush, it gets contaminated the first time you use it. Since the purpose of brushing is to remove bacteria that cause gum disease and tooth decay, it could conceivably be self-defeating to use a contaminated toothbrush—you'd be reinoculating yourself every time you brushed. But it isn't clear how much of a danger this poses in the real life of a healthy person.

How long the microbes survive on your brush varies—anaerobic germs tend to die unless they have some hollow space in the brush where they can thrive. Most plastic brushes have no such hiding places, but the handle of an electric toothbrush was found in one case to provide the right airless environment for *Porphyromonas gingivalis*, a germ that causes gum disease, to live happily. This same study found on used toothbrushes *Streptococcus mutans* and, most persistently, *Candida albicans*, a very common fungus that most people live happily with but that can cause mouth sores and vaginitis and sometimes, usually in the immunocompromised, systemic illnesses that can be fatal. Most toothbrush germs are gone within 24 hours if the toothbrush dries out.

Finding these organisms on toothbrushes shouldn't be surprising—they are among the most common germs we all live with, and unless you are elderly, pregnant, or immunocompromised, they're usually harmless. For most people, the advice to shake the water off your toothbrush and let it dry will be sufficient to prevent any serious threat. For those who want to be extra careful, soaking a toothbrush for 20 minutes in Listerine eliminates most germs the brush might harbor.

How about the bath mat? There are several studies that document the presence of bacteria and fungi in carpeting, but there isn't any evidence that their presence there causes any increase in infection rates. A clean bath mat will attain a certain level of contamination fairly quickly and stay at that level until it is washed. After washing, the bacterial count goes down and then quickly rises again. A wet mat is a better place for bacteria to grow than a dry one. Sanitizing your bath mat or any other laundry is probably not essential to good health, but if you really want to, there are ways to do it. We'll discuss them in the next chapter.

4

Whiter Whites and Brighter Colors

HEALTHY LAUNDRY

A chemist at the University of California, Davis named Gang Sun apparently felt that the smelly sock was a neglected area of research, so he set about finding a solution to this problem. According to a press release dated October 3, 2000, textiles are a great place for bacteria to grow, and bacteria and yeasts break down perspiration in your socks to create that all-too-familiar bouquet. But Mr. Sun came to the rescue by discovering a method for attaching chlorine-containing molecules called halamines to textile fibers. The chlorine kills bacteria without any unpleasant side effects—no poisonous chlorinated carbon atoms are produced. By finding this practical way of binding the halamines to cotton, Sun is now able to offer textile manufacturers a process that can easily incorporate the antismell chemical into the ordinary manufacturing process. The process can be used for single-use products such as diapers and in multiple-use products as well, because the fabric can be "recharged" by washing it in chlorine bleach. The process, says Sun, might also be used for hospital clothing and surgical scrubs.

Sun put his product to the test with a tough group: the U.C. Davis cross-country team. The coach of the team put the socks on both men and women runners and reported with great pleasure that the socks "were comfortable, not irritating, and smelt very mildly of chlorine before and after" they were used. "After an eight- or nine-mile run,"

she added, "for socks not to smell of feet is a real bonus." Who could disagree with her? Most people would probably consider it a real bonus for anything not to smell of feet at any time. The University of California has licensed the process to a firm in Seattle that plans to develop it for use not only in socks but in kitchen tools, hospital and prison uniforms, diapers, and incontinence pads as well.

It's not sweat that smells—in fact, sweat has no odor at all. But get sweat in contact with the bacteria that are your constant companions, and you wind up with a terrible smell because moist clothing is a great place for bacteria to reproduce. Of course, no one ever died from smelling socks, not even those of a cross-country team, so you may wonder about the health benefits of Mr. Sun's invention. Even when it comes to sanitizing hospital and surgical clothing, as we will see, they are probably minimal.

Perhaps even bigger offenders than feet in producing body odors are armpits. Again, it isn't the sweat that stinks. Armpits, as one scientific study puts it, are "an area of the body with exceptional odor-producing capabilities." Of course, they don't use a rude term like "armpit." They call them "axillary areas," but that doesn't make them smell any better. Their exceptional smell-producing ability comes from proteins in sweat that don't smell by themselves, but release incredible amounts of foul odor when they are used as food by bacteria. There are several types of skin glands that produce protein-laden moisture in armpits, including apocrine glands, which release part of the gland itself, along with sweat and sebaceous glands, which secrete fatty material, and there are plenty of skin bacteria to live off these nourishing secretions. These microflora include micrococci as well as the bacteria that make the strongest smell of all, the aerobic diphtheroids such as *Corynebacterium*, the bacterium that causes diphtheria, and *Propionibacterium acnes*, which can cause acne in those sensitive to it. And of course these germs get on your shirt, or your undershirt, or whatever clothing touches areas of the body where bacteria and sweat get together. So dirty laundry smells bad, too, and it's the bacteria, some of which are disease-causing under the right conditions, that are causing the odor.

In their web site under a heading called "Understanding Germs in Laundry," Procter & Gamble sums up a discussion among a group of microbiologists about disease transmission through dirty clothes. Their conclusion is that germs in laundry, insofar as they exist and are transmitted by this route, are probably not much of a threat to health, and Procter & Gamble accurately reports this. But then, ignoring the

results of this discussion among scientists, P&G goes on to eulogize the germ-fighting power of its Tide with Bleach product, with its "patented oxygen bleaching system," as "the only detergent that sanitizes laundry when a full 285mL dose is used." In other words, Tide with Bleach will sanitize your laundry (it "Kills 99.9% of Germs"), even though, as our own experts just told us, sanitizing laundry doesn't really matter.

Procter & Gamble isn't telling any lies; they're just trying to sell laundry soap. If there are people out there who want to sanitize their laundry—and there apparently are—P&G stands ready to help them. After all, if you're selling laundry soap that will "kill 99.9% of germs," it certainly can't hurt to make that part of your sales pitch.

Hospital Laundry

Dirty laundry, as you might have guessed, certainly contains plenty of germs, including many that are pathogenic. Any cloth that touches your body is likely to become contaminated with whatever organisms are on your skin—and we can leave to the imagination of anyone who has read the first few chapters of this book what can happen in the special case of underwear. But the Centers for Disease Control agrees with Procter & Gamble's panel of experts: the risk of transmitting disease by the contaminated laundry route is negligible in your home. Even in hospitals, where the risks are obviously greater, there are very few reports of diseases linked to soiled fabrics, and the overall risk of such transmissions is vanishingly small.

The clothing of health care workers, of course, presents problems not ordinarily encountered by the rest of us. Blood, skin, stool, urine, vomit, and other body fluids and tissues commonly contaminate the clothing of hospital workers, and there have been cases in which disease transmission has apparently occurred from handling these clothes. Usually the reason is that they were handled improperly—shaken, for example, causing pathogens to become airborne. Salmonella, *B. cereus*, hepatitis B virus, fungi, and scabies may all have been transmitted in this way, either by direct contact or by contact with airborne particles of lint from the laundry, even though reports of such incidents are unusual enough to warrant publication as articles in scientific journals, and their conclusions are uncertain enough that none of the authors could absolutely rule out transmission from some other source.

But the CDC doesn't like to gamble, and on the minuscule chance that infected laundry and bedding might be a source of trouble in hospitals, they do offer some thoughts about how such materials should be handled. Some of their recommendations could be adapted for home use; others require special equipment or procedures that would be impractical, if not impossible, to use at home.

That health care workers' clothes get contaminated with germs is indisputable. One study looked at those little white jackets that medical students wear and found that the sleeves and pockets were commonly contaminated. The most frequently found germs were the ones we all walk around with on our skin, including *S. aureus*, which can cause pretty awful disease if it gets into the wrong place. In general, students were more or less correct when they guessed how dirty their coats were—that is, their guesses correlated well with how contaminated they turned out to be. Nevertheless, they didn't launder them much, and a large proportion of them wore them for weeks on end no matter how dirty they were. The authors of the study concluded that it might be a good idea if the hospital would take on the burden of providing clean coats from time to time, but they didn't claim to show that anyone had caught anything from medical student coats, no matter how long they'd gone unwashed.

Washing this stuff is no simple matter. In a hospital, the laundry room has to be divided into two separate sections, one for receiving and handling dirty laundry and one for handling the material after it's clean. In order to minimize the potential for contaminating the clean area with airborne stuff from the dirty area, the air pressure in the dirty area has to be kept lower than that in the clean area. This prevents anything nasty from blowing out of the "dirty room" when you open the door. Laundry workers are supposed to wear protective gloves and clothing while handling contaminated fabrics, equipment should be maintained according to manufacturers' instructions to prevent microbial contamination, and damp laundry shouldn't be left lying in machines overnight. It used to be thought that laundry from isolation areas of the hospital (for example, wards in which TB cases are treated) should be kept separate from other laundry, but there is no evidence that this prevents disease, and the CDC now recommends that all laundry be handled in the same manner.

If you want to be really, really safe, the CDC provides detailed rules about how laundry should be handled in hospitals, and at least some of

these procedures could conceivably be instituted at home. First, a health care worker has to collect the laundry, which is not as simple as it might sound. The laundry should not be shaken or sorted at the site where it is collected. Instead, it must be placed inside a bag, sealed to prevent leakage, and then transported. This transportation can be done either by a cart or through a laundry chute, but if a chute is used, it has to be designed and maintained properly, and you can't throw laundry down it that isn't bagged properly—doing that might make infective material airborne and contaminate the laundry room at the other end of the chute.

Laundry has to be sorted before washing. In a hospital, this has to be done because it may contain "sharps"—discarded needles, lancets, or other medical waste that could carry infection—and might also contain patient belongings that could damage laundry machines. The procedure also reduces the possibility of recontamination of clean material. You would probably not be likely to find discarded medical waste in home laundry, but you might well find hard objects left in pockets or accidentally picked up with the dirty sheets and pillowcases. In hospitals, workers wear protective clothing and gloves of sufficient thickness to minimize penetration by sharp objects. It is difficult to conceive of an ordinary person, no matter how concerned about germs, getting him- or herself up in such garb in order to pop a load of clothes into the washing machine.

All these procedures, and we haven't even gotten to actually washing the stuff yet. Well, here we go. Laundry in hospitals can be disinfected and rendered "hygienically clean," although not sterile, by laundering it. That is, the material will be free of most harmful germs, but you can still find microbes on it if you look hard enough.

How do clothes get hygienically clean inside a hospital washing machine? Through a combination of factors, some of which can be duplicated in a home laundry. Just the act of washing in ordinary water removes quite a few germs. Soaps and detergents loosen soil, removing germs along with it, and some detergents have germ-killing properties as well. Hot water kills germs, as we know, but it has to be really hot. Most home water systems keep the hot water at about 140° F, but for hot water to be an effective disinfectant, it has to be raised to 170° F, and it has to stay at that temperature for at least 25 minutes. This just isn't going to happen in most homes, and even if you could manage to attain such a temperature, there are plenty of clothes you wouldn't want

to wash in water that hot. Even in hospitals, keeping water that hot may not be practical—laundries are typically the largest water users in hospitals, and heating water to such a high temperature can be a very expensive proposition, amounting to as much as 15 percent of the hospital's energy bill.

Fortunately, it isn't necessary to have water that hot, even in a hospital washing machine, if you use the right detergent and bleach combination. Chlorine bleach is a very good disinfectant, and using it will undoubtedly make your laundry less germy. It can also make a mess of your clothes. But there is an alternative method of bleaching called oxygen activated. (Remember that "patented oxygen bleaching system" that Procter & Gamble brags about in its Tide detergent?) This method doesn't involve chlorine, but is still a quite effective antimicrobial and is much easier on clothing. The temperatures reached in dryers, and when clothes are pressed, probably also contribute to the elimination of some germs.

Despite all this, there is at least one recent study that appears to demonstrate that washing surgical scrubs at home is just as healthful as having the hospital laundry go though these complicated processes. The study, published in the *American Journal of Infection Control,* showed that washing clothes at home produced no increase in surgical site infections. While the Association of Perioperative Registered Nurses opposes this, the CDC says the question is undecided. On the basis of his own results, the author concludes that there is no reason to have special laundries handle these garments.

Should you disinfect the washing machine and dryer? No. If you follow the right laundering procedures, there is absolutely no advantage in scrubbing the tubs and tumblers of these machines with antibacterials.

There are some things you can't throw into the washing machine, so dry cleaning should take care of sanitizing those, right? Probably not. Dry cleaning, a process that uses solvents such as perchloroethylene instead of water, is ineffective in reducing the numbers of bacteria and viruses on contaminated clothing. If the dry-cleaned articles are pressed with a hot enough iron, however, that does help remove some germs.

Are you enough of a mysophobe to concern yourself with the germ-carrying capacity of your mattress? Of course, mattress covers can be laundered, but how about the mattress itself? Can it get contaminated, and if it can, what should you do about it? In hospitals, particularly in burn wards, contamination of mattresses can be a problem. Wet mat-

tresses are a particularly comfortable home for colonies of bacteria, and any mattress that is wet or visibly stained should be replaced. Mattress covers should be replaced, not repaired, when torn. Pillows and pillowcases can both be laundered when soiled. We're talking about hospital procedures here, and while you could certainly implement these practices at home, they are unlikely to be a significant contributor to good health.

Germ-Fighting Pajamas?

Many consumer products are advertised as having been treated with antimicrobial chemicals. These include not only clothing, but pens, cutting boards, toys, household cleaners, hand lotions, cat litter, soap, cotton swabs, toothbrushes, and various cosmetics. There are children's pajamas, mattresses, and bed linens for sale that make such claims. The CDC is unequivocal about the advantages of using such products: "At present there is no evidence that use of these products will make consumers and patients healthier or prevent disease."

While the Food and Drug Administration is in charge of regulating personal-care products such as soaps and hand lotions, it is the Environmental Protection Agency that regulates chemicals that kill or claim to kill germs on non-living surfaces, including clothing and bedding. The law says that if a manufacturer wants to make a claim that a product "fights germs" or has "antibacterial action" or "controls fungus" or any other such health claim, that product has to be approved by the EPA. To gain that approval, the manufacturer has to apply to the EPA and provide specific data that demonstrate that the product does what he claims it does (see chapter 12 for more details on this subject). The clarity of this law, however, does not stop people from disobeying it. Therefore, the EPA urges everyone not to rely on antibacterial claims as a substitute for commonsense hygienic practices.

Moldy Laundry

Laundry rooms and wet clothes are famous locations for molds and mildew. What are these things? They're fungi, plants without chlorophyll, microscopic members of the same family of plants as mushrooms.

And there are lots of them: they make up about a quarter of the earth's biomass. The fungi can be roughly divided by their morphology into yeasts, mushrooms, and molds. Molds ("mildew" is a name commonly applied to many different molds) are the ones found in damp places—in bathrooms, cellars, laundry rooms, on rotting food, and even on wet laundry, if you leave it around long enough. Most of them come into damp places inside houses from outside, but some (*Penicillium* and *Aspergillus* are two examples) can grow inside buildings that are quite dry. Like bacteria, they're everywhere—indoors, outdoors, on plants and animals, in piles of dry leaves, in rotting organic material. They come in a variety of lovely colors—black, orange, green, brown, white. Most molds are harmless, and many are useful or essential in living systems.

But when they grow inside houses in sufficient quantity, they smell bad, very musty—take a whiff of a pile of wet laundry that's been sitting around for a few days. This is because the molds produce a large number of organic compounds that are very volatile (volatile here means not that they are explosive—they aren't!—but that they are easily evaporated into the air, where you can inhale their odor). But their smell isn't their worst characteristic. The problem with them is that when certain strains of the fungus are living on certain organic materials, they produce mycotoxins—organic compounds that can poison vertebrates, including humans. It is very difficult to know whether these mycotoxins are actually present. The toxins can be isolated in a laboratory from a sample of the mold, but even this doesn't prove that they were in the air where the sample was taken. There is one laboratory in Lenexa, Kansas, called the IBT Reference Laboratory, that has developed an isotype-specific enzyme-linked immunosorbent assay (ELISA) method to measure the presence of antibodies to the poisons in blood samples, but the Food and Drug Administration has not evaluated or approved the test. The necessary work to determine the test's sensitivity and accuracy has not been carried out, and the test is therefore not useful as a diagnostic tool. *Stachybotrys chartarum*, a greenish-black household mold less common than some others but not at all rare, has been the subject of a number of press reports that connect it to infant pulmonary hemorrhage cases, but the CDC considers these unconfirmed rumors, not scientific facts.

Still, getting these toxins in your mouth or inhaling them into your lungs can probably cause some unpleasant symptoms, even if they aren't terribly dangerous. Not all molds produce mycotoxins, and even

Common Household Molds and Their Effects

Mold species	What it lives on	What it produces	Health effects from inhalation
Alternaria alternata	Moist windowsills, walls	Allergens	Asthma, allergy
Aspergillus versicolor	Damp wood, wallpaper glue	Mycotoxins Volatile organic compounds (VOCs)	Unknown
Aspergillus fumigatus	House dust, potting soil	Allergens	Pneumonitis
Cladosporium herbarum	Moist windowsills, wood	Allergens	Asthma, allergy
Penicillium chrysogenum	Damp wallpaper, behind paint	Mycotoxins	Unknown
Penicillium expansum	Damp wallpaper	Mycotoxins	Possible kidney toxicity
Stachybotrys chartarum	Wet carpeting, gypsum board	Mycotoxins	Dermatitis, mucosal irritation, immunosuppression

Adapted from S.V. McNeel and R.A. Kreutzer, "Fungi and Indoor Air Quality." California Department of Health Services, Environmental Health Investigations Branch. *Health & Environment Digest*, May/June 1996; 10: 9–12.

those that do produce them do so only under certain physical conditions. But poisons called aflatoxins have been isolated in house dust containing *Aspergillus flavus* and some called trichothenes have been found in *Fusarium, Cephalosporium, Stachybotrys,* and *Trichoderma,* all common species of household molds.

No one knows exactly how these toxins make us sick, but they do cause a variety of unpleasant symptoms. Repeated exposure to airborne mycotoxins can cause nose, throat, and eye irritations, allergy, aggravation of asthma, and, particularly in immunocompromised people, pneumonitis, a dangerous inflammation of the lungs. If you find mold or mildew in your laundry room (or anywhere else), it's a good idea to get rid of it and then eliminate the conditions that allow it to grow.

The easiest and cheapest way to remove mold is with a chlorine bleach solution—one and a half cups to a gallon of water. Just make sure the area is well ventilated, since chlorine itself is not especially healthy to inhale. Mold is easiest to clean up when it is damp, and you should avoid touching it. Wear rubber gloves. Dried-out mold is more likely to release spores, and if you're cleaning that up you should wear a dust mask. Moldy garbage should be put into plastic bags and tied tightly before being discarded.

Then clean up the obvious sources of unwanted moisture—leaky pipes, windows, roofs, and so on. Rain and irrigation water should be directed away from exterior walls. Ventilation systems and dehumidifiers can be helpful in places where the source of moisture can't easily be eliminated, and there are mold inhibitors that can be added to paints. Drying clothes indoors or putting carpeting in laundry rooms or bathrooms are invitations to mildew. If you are tempted to leave wet laundry in the washing machine or dryer, resist. And finally, stop worrying. Toxic molds are commonly found in houses, and although they can pose significant dangers to those who are immunocompromised, they almost never cause serious illness in otherwise healthy people.

In short, there are some things you can do to "sanitize" your laundry and your laundry room, mostly by performing home versions of hospital procedures. But the contribution of such efforts to your physical health is dubious at best. In weighing the costs against the benefits, there are undoubtedly better ways to protect your health than by trying to produce "99.9% germ-free" laundry.

5

Clean Up That Room

KIDS AND MICROBES

Kids are unquestionably messy, and the younger they are the messier they seem to be. Growing up involves an incessant series of encounters with germs of all kinds. Babies begin by fouling themselves with germ-laden feces, then as soon as they're able to they start putting everything they can lift into their mouths. Next they get repeated infections—runny noses, ear infections, fevers, rashes, you name it. And just when you get past this, they become preadolescents and teenagers who for obscure reasons won't bathe as often as you'd like them to, or, alternatively, bathe so much that you're surprised they haven't washed their skin right off. Then they go off to college and are exposed to diseases you don't even want to talk about. How do they manage to survive the onslaught?

At least in well-off industrialized countries, most kids flourish despite their constant exposure to pathogenic microbes and their dubious personal hygiene habits. Since parents' everyday attempts to protect their children from germs are largely ineffective—try explaining to a two-year-old why he shouldn't put dirty toys into his mouth—there must be something else at work here. In fact, there are several things at work, some of which are in the control of caregivers, and some of which are not.

A Ton of Prevention

Viruses are parasitic. They can only live and reproduce inside the cells of other animals or plants—in scientific terms, they are called obligate

intracellular molecular parasites. They are obligate because they can only survive in one environment (if a germ can survive in more than one environment, it is called facultative); they are intracellular because they must live inside cells; they are molecular parasites because they are the size of molecules and can only live in a parasitic relationship to another species. They are very tiny—even the largest of them, the pox virus for example, is about one-fifth the size of an *E. coli* bacterium, and the smaller ones, among which is the polio virus, are one tenth the size of a pox virus. They cannot be seen with a light microscope. You have to use an electron microscope, X rays, or nuclear magnetic resonance in order to see their structure.

It's hard to say whether viruses are animals or plants, and in some senses they are just barely alive at all. They consist of nucleic acid, either DNA or RNA, and their sex life, such as it is, consists of their genomes directing the synthesis of their components inside the proper host cell. These components are assembled into new viruses that can then infect other cells. There are many different species, classified by their characteristics. Some contain DNA, others RNA. They vary in shape and size. Some have a lipid "envelope" derived from the membrane of the host cell, as in the case of the virus that causes herpes; others, like polio and the papillomaviruses (the ones that cause warts) lack this envelope. The capsid, the protective shell that surrounds the viral genome, varies in size and shape from one virus to another.

More than a thousand years ago, people noticed that once you recovered from certain illnesses, such as smallpox, you could never get them again. Since getting the disease seemed to protect people against getting it again, they began deliberately infecting people with small amounts of material from smallpox pustules—scratching their arms with it to infect them with a mild version of the disease so that they would be protected against the more serious version. This worked well in most cases, but in some it led to bad skin lesions all over the body, and about 1 percent of people who underwent this procedure (called "variolation") died from smallpox. This is not the sort of treatment that would today survive review by the Food and Drug Administration, but for many people it was unquestionably a lifesaver.

Then, about 200 years ago, an English physician named William Jenner noticed that milkmaids, who were commonly infected with cowpox, a mild disease that leaves no scars, were immune to smallpox. (Thus their reputation for having lovely complexions unscarred by smallpox.) So Jenner decided to deliberately infect people with cowpox,

risking a mild illness in the hope that it would protect them against smallpox, an often fatal disease. Remarkably, it worked, and the custom of vaccination (Latin *vacca* = cow) was born.

By the 1890s, Louis Pasteur had figured out how to create an "attenuated" or weakened virus that was strong enough to elicit a protective immune response, but not strong enough to cause disease. This was an important step forward in vaccine safety.

Today there are three different kinds of vaccines: live attenuated virus vaccines, inactivated virus vaccines, and acellular vaccines. The measles vaccine is an example of the first kind—it's a vaccine made from a live attenuated (weakened) form of the disease-causing virus. Live vaccines are extremely effective—usually one or two doses are enough to protect you for life without ever having to get a booster shot. Rubella and mumps vaccines are also live virus. So are the varicella (chicken pox) vaccine and the oral polio vaccine. The second category is the inactivated virus vaccines. The polio vaccine used in the United States is of this type.* The vaccines for diphtheria and tetanus are also inactivated vaccines, but for these diseases and others booster vaccines are necessary throughout life.

Another category is the acellular vaccines. The vaccine for pertussis, for example, contains not the bacterium that causes whooping cough, but only purified antigenic components of the pertussis bacterium—that is, the part of the cell that produces immunity. Finally, there are vaccines like the one for hepatitis B, which is chemically engineered by recombinant DNA techniques. There are now approved vaccines against 21 different diseases (some diseases can be prevented with more than one vaccine). Eleven of these are recommended for all children; the other ten are used only for populations at risk because of age, residence area, medical condition, or risk behaviors. The reasons why only 11 of the 21 vaccines are recommended for everyone are complex and quite interesting, and we'll come back to them later in this chapter when we discuss the problem of the vaccine for meningococcal meningitis. Here are the vaccines currently available in the United States:

*The oral polio vaccine (OPV) is used in most parts of the world. As of January 2000, however, only the inactivated virus vaccine (IPV) is used in the United States. Both forms give immunity to polio, but the OPV is somewhat more effective in preventing spread to other (unvaccinated) people. The problem with it is that OPV itself causes polio in one in 2.4 million vaccinations. The risk/benefit ratio in the United States, where polio is extremely rare, has been judged lower with IPV.

Licensed Vaccines and Toxoids Available in the United States

Vaccine	Type	Route
Adenovirus	Live virus	Oral
Anthrax	Inactivated bacteria	Subcutaneous
Bacillus of Calmette and Guérin	Live bacteria	Intradermal/ percutaneous
Cholera	Inactivated bacteria	Subcutaneous or intradermal
Diphtheria-tetanus-pertussis	Inactivated bacteria	Intramuscular
Haemophilus influenzae type b conjugate	Bacterial polysaccharide conjugated to protein	Intramuscular
Hepatitis B	Inactive viral antigen	Intramuscular
Influenza	Inactivated virus or split virus	Intramuscular
Japanese encephalitis	Inactivated virus	Subcutaneous
Measles	Live virus	Subcutaneous
Measles-mumps-rubella	Live virus	Subcutaneous
Meningococcal	Bacterial polysaccharides	Subcutaneous
Mumps	Live virus	Subcutaneous
Pertussis	Inactivated whole bacteria	Intramuscular
Plague	Inactivated bacteria	Intramuscular
Pneumococcal	Bacterial polysaccharides	Intramuscular or subcutaneous
Polio	Inactivated virus	Subcutaneous
Polio	Live viruses	Oral
Rabies	Inactivated virus	Intramuscular
Rubella	Live virus	Subcutaneous
Tetanus	Inactivated toxin	Intramuscular
Tetanus-diphtheria	Inactivated toxins	Intramuscular
Typhoid (parenteral)	Inactivated bacteria	Subcutaneous
Typhoid	Live bacteria	Oral
Varicella	Live virus	Subcutaneous
Yellow fever	Live virus	Subcutaneous

CDC, *Morbidity and Mortality Weekly Report*, January 28, 1994, vol. 43, no. RR-1.

The most important thing to do to prevent infectious disease in children is to get them vaccinated, do it on time, and keep the vaccinations current. There are now safe and effective childhood vaccines that will prevent diphtheria, *Haemophilus influenzae* Type B (Hib), hepatitis A, hepatitis B, measles, mumps, pertussis (whooping cough), polio, rubella (German measles), tetanus (lockjaw), and varicella (chicken pox). These diseases are now very uncommon in the United States, thanks to an effective vaccination system. But they flourish elsewhere in the world, in places only a plane ride away, and that makes efficient vaccination essential even in a place where these diseases have been largely eliminated.

The first of the following two tables offers a description of the disease and vaccine for each of the vaccine preventable diseases. The second table shows the recommended schedule of vaccination. Both are adapted from CDC publications.

Vaccine-Preventable Diseases of Childhood

Disease	Description	Symptoms	Complications	How You Get It	Vaccine
Diphtheria	Bacterial respiratory illness	Gradual onset of a sore throat and low-grade fever	Airway obstruction, coma, and death if untreated	Coughing, sneezing	Diphtheria toxoid (contained in diphtheria-tetanus-pertusis [DTP], diptheria-tetanus-acellular pertussis [Dta P], diptheria-tetanus [DT] or tetanus toxoid [Td] vaccines)
Haemophilus influenza type b (Hib)	Bacterial. Primarily in infants	Stiff neck, difficulty in breathing	Death or permanent brain damage	Coughing, sneezing	Hib vaccine
Hepatitis A	Viral liver disease	Sometimes none; sometimes stomachache, loss of appetite, jaundice	Infants and children may be asymptom-atic. Often not noticed until adult caregiver becomes ill	Fecal-oral route	Hepatitis A vaccine
Hepatitis B	Viral liver disease	Sometimes none; sometimes stomachache, loss of appetite, jaundice	Lifelong liver disease, liver cancer	Blood contact or sexual contact	Hepatitis B vaccine
Measles	Viral respiratory disease	Rash, fever, cough, runny nose lasting one week	Diarrhea, ear infection, pneumonia, encephalitis, seizures, death	Coughing, sneezing. Extremely contagious	Measles vaccine (contained in measles-mumps-rubella [MMR], measles-rubella [MR] and measles vaccines)
Mumps	Generalized viral disease	Fever, headache, muscle ache, swelling of salivary glands	Meningitis, inflammation of the testes, ovaries, pancreas, deafness.	Coughing, sneezing	Mumps vaccine (contained in MMR)

(continued)

Vaccine-Preventable Diseases of Childhood *(continued)*

Disease	Description	Symptoms	Complications	How You Get It	Vaccine
Pertussis (whooping cough)	Bacterial respiratory disease	Severe spasms of coughing	Pneumonia, encephalitis, seizures, death in infants	Coughing, sneezing, very contagious	Pertussis vaccine (contained in DTP and DtaP)
Polio	Viral disease of nervous and lymphatic systems	Fever, sore throat, nausea, headaches, muscle stiffness in neck, back, and legs	Paralysis, permanent disability, death	Contact with an infected person	Polio vaccine inactivated polio virus (IPV)
Rubella (German measles)	Viral respiratory disease	Rash, fever, usually mild in children	Birth defects if acquired by a pregnant woman. Infant may develop deafness, cataracts, heart defects, mental retardation, liver and spleen damage	Coughing, sneezing	Rubella vaccine (contained in MMR)
Tetanus	Bacterial disease. Toxin-mediated damage to the nervous system	Early: lockjaw, stiffness, difficulty swallowing. Later: fever and muscle spasms	Death in 33% of cases, especially people over 50	Enters through break in the skin	Tetanus toxoid (contained in DTP, DT, DtaP and Td vaccines)
Varicella (chicken pox)	Virus of the herpes family	Skin rash of blister-like lesions	Bacterial infection of the skin; pneumonia; more severe in adults	Coughing and sneezing; highly contagious	Varicella vaccine
Pneumococcal disease or infection	Bacteria	Lung infection	Lung disease, meningitis, bacteremia, circulatory collapse, death rate of 18% in patients over 12 years old	Fever, chills, chest pain, cough, rusty sputum, shortness of breath	Pneumococcal conjugate vaccine (PCV) for children up to 9. Pneumovax for the elderly

Recommended Childhood Immunization Schedule*
United States, 2002

Age ▸ Vaccine ▾	Birth	1 mo	2 mos	4 mos	6 mos	12 mos	15 mos	18 mos	24 mos	4-6 yrs	11-12 yrs	13-18 yrs
Hepatitis B[1]	Hep B #1	only if mother HBsAg(neg.)										
		Hep B #2			Hep B #3					Hep B series		
Diphtheria, Tetanus, Pertussis[2]			DTaP	DTaP	DTaP		DTaP			DTaP	Td	
Haemophilus influenzae Type b[3]			Hib	Hib	Hib	Hib						
Inactivated Polio[4]			IPV	IPV		IPV				IPV		
Measles, Mumps, Rubella[5]						MMR #1				MMR #2	MMR #2	
Varicella[6]						Varicella				Varicella		
Pneumococcal[7]			PCV	PCV	PCV	PCV	PCV		PCV	PPV		
Hepatitis A[8]										Hepatitis A series		
Influenza[9]								Influenza (yearly)				

Vaccines below this line are for selected populations

*Indicates the recommended ages for routine administration of currently licensed childhood vaccines, as of December 1, 2001, for children through age 18 years. Any dose not given at the recommended age should be given at any subsequent visit when indicated and feasible. Indicates age groups that warrant special effort to administer those vaccines not given previously. Additional vaccines may be licensed and recommended during the year. Licensed combination vaccines may be used whenever any components of the combination are indicated and the vaccine's other components are not constrained. Providers should consult the manufacturers' package inserts for detailed recommendations.

Table 4 CDC: Advisory Committee on Immunization Practices.

Footnotes begin on next page.

1. Hepatitis B vaccine (Hep B). All infants should receive the first dose of hepatitis B vaccine soon after birth and before hospital discharge; the first dose may also be given by age 2 months if the infant's mother is HBsAg-negative. Only monovalent hepatitis B vaccine can be used for the birth dose. Monovalent or combination vaccine containing Hep B may be used to complete the series; four doses of vaccine may be administered if combination vaccine is used. The second dose should be given at least 4 weeks after the first dose, except for Hib-containing vaccine, which cannot be administered before age 6 weeks. The third dose should be given at least 16 weeks after the first dose and at least 8 weeks after the second dose. The last dose in the vaccination series (third or fourth dose) should not be administered before age 6 months. Infants born to HBsAg-positive mothers should receive hepatitis B vaccine and 0.5 ml hepatitis B immune globulin (HBIG) within 12 hours of birth at separate sites. The second dose is recommended at age 1–2 months, and the vaccination series should be completed (third or fourth dose) at age 6 months. Infants born to mothers whose HBsAg status is unknown should receive the first dose of the hepatitis B vaccine series within 12 hours of birth. Maternal blood should be drawn at the time of delivery to determine the mother's HBsAg status; if the HbsAg test is positive, the infant should receive HBIG as soon as possible (no later than age 1 week).

2. Diphtheria and tetanus toxoids and acellular pertussis vaccine (DTaP). The fourth dose of DTaP may be administered as early as age 12 months, provided that 6 months have elapsed since the third dose and the child is unlikely to return at age 15–18 months. **Tetanus and diphtheria toxoids (Td)** is recommended at age 11–12 years if at least 5 years have elapsed since the last dose of tetanus and diphtheria toxoid-containing vaccine. Subsequent routine Td boosters are recommended every 10 years.

3. Haemophilus influenzae type b (Hib) conjugate vaccine. Three Hib conjugate vaccines are licensed for infant use. If PRP-OMP (PedvaxHIB or ComVax [Merck]) is administered at age 2 and 4 months, a dose at age 6 months is not required. DTaP/Hib combination products should not be used for primary immunization in infants at age 2, 4, or 6 months but can be used as boosters following any Hib vaccine.

4. Inactivated poliovirus vaccine (IPV). An all-IPV schedule is recommended for routine childhood poliovirus vaccination in the United States. All children should receive 4 doses of IPV at age 2 months, 4 months, 6–18 months, and 4–16 years.

5. Measles, mumps, and rubella vaccine (MMR). The second dose of MMR is recommended routinely at age 4–6 years but may be administered during any visit, provided at least 4 weeks have elapsed since the first dose and that both doses are administered beginning at or after age 12 months. Those who have not previously received the second dose should complete the schedule by age 11–12 years.

6. Varicella vaccine. Varicella vaccine is recommended at any visit, at or after age 12 months for susceptible children (i.e., those who lack a reliable history of chicken pox). Susceptible persons aged >13 years should receive 2 doses, given at least 4 weeks apart.

7. Pneumococcal vaccine. The heptavalent **pneumococcal conjugate vaccine (PCV)** is recommended for all children aged 2–23 months and for certain children aged 24–59 months. **Pneumococcal polysaccharide vaccine (PPV)** is recommended in addition to PCV for certain high-risk groups. See *MMWR* 2000; 49 (RR-9); 1–37.

8. Hepatitis A vaccine. Hepatitis A vaccine is recommended for use in selected states and regions and for certain high-risk groups. Consult local public health authority and *MMWR* 1999; 48 (RR-12); 1–37.

9. Influenza vaccine. Influenza vaccine is recommended annually for children aged >6 months with certain risk factors (including but not limited to asthma, cardiac disease, sickle cell disease, HIV, and diabetes; see *MMWR* 2001; 50 [RR-4]; 1–44), and can be administered to all others wishing to obtain immunity. Children aged <12 years should receive vaccine in a dosage appropriate for their age (0.25 ml if 6–35 months or 0.5 ml if >3 years). Children aged <8 years who are receiving influenza vaccine for the first time should receive 2 doses separated by at least 4 weeks.

Unlike some of the things we've discussed in this book (like sanitizing your laundry or disinfecting your toothbrush, for example), getting these vaccines, and making sure that your kids get them, is unquestionably essential for your continuing good health. There are some fears about side effects of various vaccines, but the latest studies show that this risk is dwarfed by the dangers of getting any of the vaccine-preventable diseases. Often the study that shows a connection between a given childhood illness and a vaccine gets much more publicity than the numerous studies that contradict those findings. Widely publicized assertions that the MMR vaccine is associated with increased risk for autism (based on a small study that included 12 children) have been roundly refuted, but many parents are aware only of the original unscientific assertion, not the contradicting studies. The Centers for Disease Control conducts ongoing studies of vaccine safety in a project called the Vaccine Safety Datalink (VSD). VSD has data on more than 6 million people, and their medical records are examined continuously for adverse events relating to vaccination. At any given time, the CDC is conducting dozens of studies of various vaccines and publishing the results of their findings. In other words, these recommendations are not made casually or carelessly. They are based on the latest scientific knowledge and are modified as new knowledge is acquired.

In addition, the CDC runs a program called the Vaccine Adverse Event Reporting System (VAERS), and anyone can report what he or she believes to be an adverse reaction to a vaccine. Most reports to VAERS come from manufacturers of vaccines and health care professionals, but about 3 percent come from patients or parents of patients. There is a form for making this report that can be obtained by calling 1-800-822-7967 or by downloading it from the Internet at the VAERS web site, www.vaers.org. It's preaddressed and postage free.

Because giving an infant vaccines is a common event, and for parents a memorable one, there may be a tendency to attribute any illness that follows the event to the vaccine. But that an illness follows a vaccination is not proof that the vaccination caused the illness. Many illnesses happen by chance, and it is the purpose of scientific studies of vaccination to distinguish between random events that follow a vaccine and events that are caused by a vaccine.

Vaccine Risk, and How to Estimate It

The Internet can be a significant source of misinformation about vaccines. There are numerous sites that gather scare stories about vaccination based on little more than anecdotal reports from parents, and many chat rooms discussing vaccination in hysterical voices and without benefit of scientific information. This doesn't mean that vaccines are without dangers—there are dangers, but the dangers must be understood scientifically, and especially in comparison to the risks involved in getting the diseases these vaccines prevent. When looking at an Internet site, ask yourself some questions: Who is the owner or sponsor of this site? Does it contain references to scientific articles published in well-known journals? Are the studies it cites anecdotes of individual cases, or are they scientific studies that require large numbers of cases and scientific controls? Is the site endorsed by a university or a professional organization? Which ones? Are the sources of information named, or are assertions attributed to "renowned scientists" or "noted researchers"? Is an assertion too good ("Eating banana peels cures AIDS!") or too absolutely appalling ("Millions die every week from flu vaccines!") to be true? Does the site reveal the limitations of its information? In other words, is this junk science or real science?

How can you compare the risk of a vaccine against the risk of getting the disease? Only with statistical information based on large studies. Measles, for example, is a highly contagious disease that leads to pneumonia in 1 of 20 cases, to encephalitis (inflammation of the brain) in 1 in 2,000 cases, and death in 1 in 3,000 cases; mumps causes encephalitis in 1 of 300 victims; rubella infection in pregnant mothers causes birth defects one in every four times. The only risk from the MMR (measles, mumps, rubella) vaccine, on the other hand, is getting encephalitis or a severe allergic reaction, which happens one time in

every 1 million vaccinations. For DTP vaccine (diphtheria, tetanus, and pertussis) the story is much the same: the risk of dying from diphtheria is 1 in 20. The only risk to an infant from the diphtheria vaccine is continuous crying, then full recovery, which happens once in every 100 injections. Tetanus causes death in 3 of 100 of its victims; the tetanus vaccine causes convulsions or shock and then full recovery once in every 1,750 injections. Whooping cough causes 1 death in 200, but no one has ever been proven to have died from a pertussis vaccination.

Vaccination rates in the United States are very high. Because states require immunizations before kids are allowed to enroll in school, about 95 percent of kids are protected against vaccine-preventable diseases by the time they are six years old. The CDC does not, however, collect data on whether the vaccinations were given at the appropriate times indicated in the table on page 77.

What about the 5 percent of children who don't get vaccinated? They get sick at higher rates, and they make other kids sick, even ones who have been vaccinated. About half of all states allow exemptions from vaccination for religious, medical, or philosophical reasons. Colorado is one of them, and in 1998, 2 percent of Colorado kids were given exemptions. The CDC studied the results, analyzing data collected between 1987 and 1998. The risk of measles among the unvaccinated was 22.2 times as high, and the risk of whooping cough was 5.9 times as high. Moreover, in counties with the highest rates of exemptions, the measles rate was 50 percent higher among vaccinated kids than in the rest of the state. Clearly, failing to vaccinate a small number of children puts a large majority at risk for illness.

All the recommended vaccines have proven safe and effective over many years of use. But sometimes a vaccine doesn't do what it's supposed to do. Such was the case with the vaccine for rotavirus.

By the time they are five years old, almost all children, in industrialized countries and elsewhere, have had the illness caused by this virus. Most recover without aftereffects. But it causes diarrhea, which in severe cases can be fatal. It kills approximately 40 children a year in the United States and hundreds of thousands worldwide. Rotavirus is the most common cause of severe gastroenteritis in infants and young children in the United States, and is a common cause of emergency room visits, outpatient clinic visits, and hospitalization—55,000 kids a year are hospitalized with it. Rotavirus diarrhea costs $264 million a year in

Figure 5 A transmission electron micrograph of rotavirus. There's no defense against this one. The vaccine for it isn't made anymore, and almost all kids are going to get sick. A small percentage of them will wind up in the hospital getting intravenous rehydration. Its wheel shape gives it its name. *CDC. Photograph by Dr. Erskine Palmer.*

direct medical costs, and more than $1 billion in societal costs attributable mainly to loss of work time among parents and caregivers. An affordable vaccine to prevent the disease is obviously a desirable alternative, and several rotavirus vaccines have been developed. One of these, an oral, live, rhesus-based preparation manufactured by Wyeth Laboratories, was approved for use by the Advisory Committee on Immunization Practices (ACIP) of the CDC in March 1999.

Approval of the vaccine came only after extensive studies of its safety and effectiveness. In the notification recommending the use of the vaccine and specifying the dosages and their timing, the CDC made note of several side effects, but none were deemed sufficiently serious to reject use of the vaccine. Among these side effects was a condition called intussusception, in which one part of the intestine becomes twisted and pushed into an adjoining segment, causing bowel obstruction and acute abdominal symptoms. Whether they have had the vaccine or not, the condition is seen most commonly in children. The researchers noted 5 cases of intussusception in 10,054 recipients of the vaccine, and 1 case in 4,633 children who had received placebos. This is not a statistically significant difference—that is, it is not enough to conclude that the vaccine is the cause of intussusception. The CDC did, however, warn that although the connection did not seem to be causal, it was nevertheless something that would have to be looked for during postlicensure surveillance. It is during this final period of examination of a drug, when very large numbers of people are using it, that appar-

ently rare side effects can be shown to be more common than clinical trials of the medicine revealed. This is exactly what happened in the case of Rotashield®, the brand name given by Wyeth to its rhesus rotavirus vaccine.

By July 1999, the CDC had suspended the use of the vaccine to review data indicating a strong association between its use and the occurrence of intussusception. This review confirmed the association, and the vaccine was withdrawn from the market in October. This led some people to believe that the CDC and the ACIP must have made some kind of mistake. But they had not. In fact, the procedure worked exactly as it is designed to work with any new drug.

There is currently no vaccine to prevent rotavirus infection, and if you have an infant you can be almost certain that at some time he or she will be a victim of the disease. Rotavirus, as its name suggests (Latin *rota* = wheel), looks like a wheel under an electron microscope. The disease is highly contagious, and a child may be contagious before the diarrhea begins and for a few days after it ends. It is usually transmitted by the fecal-oral route, but unlike most viruses it is fairly stable on environmental surfaces and can be transmitted by contact with contaminated surfaces as well as in water or food. The virus thrives from November to April. Most children are infected by the time they are two years old, and although the disease can occur in adults, it is almost always mild and transitory. In children, the illness, whose symptoms are vomiting and watery diarrhea, and usually fever and abdominal pain as well, lasts between three days and a week. After they've had it once, there is some immunity—subsequent infections tend to be much milder than the first one.

There is no treatment except to provide oral rehydration therapy to counteract the effects of the diarrhea and prevent dehydration. About one out of every 40 kids with the disease will wind up hospitalized and on intravenous fluids. The diagnosis is usually made without a test to prove that rotavirus is the culprit, but there is a lab procedure that can detect antigens in a stool sample. Because it's so contagious, it's important to follow the rules: keep children or adults with diarrhea separated from other kids; wash your hands often, and make sure everyone else in the family does, too, especially before meals and after changing a diaper or helping a kid use the toilet; use disposable paper towels for hand drying; keep toys, changing tables, bathrooms, and kitchen as clean as possible. A recent British study suggests that bottle-feeding

(with or without breast-feeding) is a risk factor for rotavirus infection, and that breast-feeding provides some protection against it. But the fact is that breast-fed or not, most kids will eventually get the disease.

Day Care Center Threat

Rotavirus is a good example of the kind of disease that is easily transmitted in child care settings. There are others as well, most of them considerably less serious in their effects than rotavirus, but nevertheless things to watch out for. Among them is hepatitis A (HAV). Although there is a vaccine for it, hepatitis A remains a threat in day care centers. Hepatitis A is a very tough virus, and the usual route of transmission is fecal-oral. It can last on environmental surfaces for more than a month, and to kill it with heat, you have to cook food to 185°F for at least a minute. Surfaces can be disinfected with a 1-to-100 solution of household bleach. No one is quite sure why, but the highest rates of HAV occur in western and southwestern states. Eleven states in the United States constitute 22 percent of the country's population: California, Alaska, Washington, Oregon, New Mexico, Utah, Arizona, Oklahoma, South Dakota, Idaho, and Nevada. But those same states are home to 50 percent of the HAV cases.

Interestingly, though, the threat is not that great for the kids themselves. Most kids who get hepatitis A have symptoms so mild that no one pays much attention to them, or they have no symptoms at all. The problem is that it is transmitted by toddlers to adults and older children, and it can make these older people very sick indeed. The very young kids serve as what epidemiologists call the reservoir of disease. The results of this transmission are serious: between 11 and 22 percent of people with the illness wind up in the hospital, and about 100 people a year die from liver complications of HAV. The fatality rates among people over 50 are the highest.

Since hepatitis A is caused mainly by contact with the feces of an infected person, most people would probably guess that being a sewage worker would put you at much higher risk for HAV than being a day care worker. But in fact, no work-related instances of HAV transmission have ever been reported among sewage workers in the United States, while outbreaks traced to day care centers have been recognized for at least the past 30 years. The cause is just what you're thinking:

poor hygiene among little kids and poor handling of dirty diapers. In day care centers that have toilet training as an admission requirement, HAV outbreaks are rare. Because HAV has few symptoms in kids, the way the outbreak is usually noticed is that a bunch of adults, usually parents, get ill. For some reason, staff at day care centers don't get ill at the same high rates as parents do.

The Centers for Disease Control and the Advisory Committee on Immunization Practices recommends hepatitis A vaccination only for high-risk groups and in high-risk geographical areas. There is currently no approved vaccination formulation or schedule for children under two years old. When a vaccine can be used for children under two, the elimination of transmission of HAV will be possible.

More Day Care Distress

Shigellosis (See chapter 2) is another disease kids risk contracting in day care centers and passing on to their families. In 1991 there was an outbreak of *Shigella sonnei* infection in Lexington-Fayette County, Kentucky, that involved 14 licensed day care centers, making a lot of people sick but—every cloud has a silver lining—giving researchers a chance to test some epidemiological hypotheses. They discovered some useful information. Day care centers that had outbreaks were more likely to have a food handler who also changed diapers and provided transportation for children from their homes to the center. These centers also had a higher toddler-to-toilet ratio, so there were more kids using the same bathroom. In most families with shigellosis, the first person with diarrhea was a child younger than six. And, perhaps most interesting, the community-wide promotion of hand washing was identified as the most likely factor in getting the outbreak under control.

A Disease Your Kid Will Almost Certainly Get

The most common cause of serious respiratory tract infections—pneumonia and bronchiolitis (infection of the small vessels in the lungs called bronchioles)—in infants and children is a germ called respiratory syncytial virus, or RSV. This is another very common virus, so common that by the age of three almost every kid has been infected with it.

In child care settings, almost 100 percent of children get RSV before they are a year old. It's easily spread by coughing, and even by touching a surface that an infected person has touched or coughed on. Kids are infectious before any symptoms appear and can remain so for up to three weeks after symptoms have disappeared.

Usually it looks like a cold and comes and goes after a week or two without much suffering. But sometimes an infant experiencing a first case of RSV gets so sick that he has to be hospitalized to treat a severe infection of the lower respiratory tract that causes respiratory distress. This is a winter virus, with outbreaks that peak in February, and while it happens everywhere, it's particularly severe among Alaskan Native infants. No one knows of any risk factors for serious RSV disease—that is, there's no way to predict which kids are likely to get seriously ill from it—so it's hard to recommend any particular practice that will prevent it from happening, and there is no vaccine against it. Premature babies or those with lung abnormalities or compromised immune systems have at times been given immune globulin with RSV antibodies in it and ribavirin, an antiviral drug often used in the treatment of HIV infection, but this is not done with otherwise healthy infants, and there is some evidence suggesting that ribavirin may do more harm than good. It is known that certain environmental conditions, such as the passive inhalation of smoke from tobacco or wood-burning fireplaces, increase the risk of respiratory infection in general. Once again, the most effective way we know to prevent the spread of this disease is by washing hands often and thoroughly. Disposing of tissues used by infected kids and the careful disinfection of environmental surfaces is also important in preventing larger outbreaks.

There is one other very controversial medicine for RSV—it doesn't cure it, but it can make symptoms less severe, which is important in vulnerable kids. It is called palivizumab and is sold under the brand name Synagis by MedImmune, a small company in Maryland. It is a humanized monoclonal antibody against RSV produced by recombinant DNA technology and accounts for about 79 percent of MedImmune's sales. Aside from the fact that it is not 100 percent effective, the big problem with the drug is that it requires five monthly shots, and each one can cost as much as $900. Naturally, this causes some considerable discussion about who should use it and who should pay for it. MedImmune has undertaken an extremely aggressive consumer advertising campaign aimed directly at the fears of parents. One

of their ads shows a picture of a doctor putting an infant on oxygen, with a caption that reads, "If you knew what RSV could do to your precious baby, it would take your breath away," and ends with this ominous sentence: "If your baby was born early, call your pediatrician now—before it's too late." This is the kind of advertising designed to make parents demand a drug and reach for their wallets—whatever the scientific studies say. While MedImmune would like to see its product used for all babies born five weeks or more prematurely, the most recent studies conclude that Synagis should only be used for infants with active chronic lung disease or ones born prematurely with multiple risk factors for RSV. While continuing their consumer advertising, MedImmune has hired 100 extra salespeople to visit doctors, in addition to 500 salespeople employed by Abbott Laboratories, the company that is helping them market the drug. In 2000, sales of Synagis increased by 46 percent over 1999. No one doubts that the drug can be effective in cases where it is needed. Exactly who needs it, however, remains in dispute.

Second only to RSV as a cause of lower respiratory illness in kids is human parainfluenza virus (HPIV). There are four different kinds of HPIV, numbered 1 through 4, and they have slightly varying symptoms. HPIV-4 is the least likely to cause severe disease. HPIV-1 is the leading cause of croup (laryngotracheobronchitis), which can also be caused by HPIV-2. HPIV-3 usually causes bronchiolitis and pneumonia. These viruses are transmitted from person to person, and they last only a few hours on environmental surfaces. Plain soap and water inactivate these germs, and there is no need to repeat the advice about hand washing.

That Kid Gave It to My Kid— Or Maybe the Other Way Around

E. coli O157:H7 (see chapter 2) is a threat with young children not only because it can make them quite ill, but also because they can shed it in their feces for up to two weeks after their illness resolves, giving them plenty of opportunity to pass the disease on to others. Older children don't usually carry the organism without having the symptoms of the disease. Since it's impossible to know which kids are still shedding the virus in their feces, *E. coli* can easily be spread around a day care center even when there isn't anyone who seems ill.

While *salmonella* and *campylobacter* infections have declined in recent years thanks to the introduction of new food safety regulations for meat and poultry and improved agricultural practices for produce, the incidence of *E. coli* O157:H7 infections has not declined. The only known person-to-person transmission of *E. coli* occurs from child to child in day care centers. If *E. coli* is to be controlled we have to make sure that everyone in day care centers washes his hands after using the toilet, and that caregivers wash the hands of kids who are too little to do it themselves. It should be unnecessary to add that adults must wash their hands after they change a child's diaper and after they help a toddler use the toilet.

Maintaining hygiene in a place like a day care center is difficult. There are lots of messy kids in such places, and the more there are and the fewer adults to supervise them, the more opportunities there are to pass germs around. Antibiotics are not as effective as they once were. We need larger and larger doses of antibiotics to treat illnesses, and some germs have become partly or completely resistant to the drugs we have. The reasons for this are complex, but one important reason is the overuse and misuse of antibiotics. Antibiotics kill germs, but germs evolve rapidly, and small changes in their genetic makeup can make them resistant to drugs. When an antibiotic is used, it kills the germs most susceptible, but leaves resistant strains to reproduce, increasing the percentage of resistant bacteria and limiting the effectiveness of the antibiotics.

This is particularly troublesome in day care centers, where resistant strains of many pathogenic microbes flourish. There are reports from all over the world—the United States, Portugal, Taiwan, Australia, Sweden, Canada, Israel, Bulgaria, Romania, and many other countries—of resistant strains of streptococcus and pneumococcus in day care centers.

Everywhere it is studied, it is clear that children in day care suffer more infections, and more antibiotic-resistant infections, than those in the same age group kept at home. But it is also true that these kids develop immunity to many illnesses sooner.

Killer Staph

While resistant strains of the potentially deadly *S. aureus* bacterium have been known for some years to occur in hospitals, it is only recently

that these strains have made their way into the community at large. In 1997, a seven-year-old girl from an urban area of Minnesota was admitted to a hospital with a temperature of 103 and pain in her right groin. Her hip joint was found to be infected and was surgically drained. She was treated with cefazolin, a cephalosporin antibiotic of the same molecular type as penicillin, but effective against more bacteria. It didn't make her any better. When they cultured her blood and joint fluid, the doctors found methicillin-resistant *Staphylococcus aureus* (MRSA), and they started her on vancomycin, the one drug that can be effective in combating this microbe. But it was too late: her hip was still infected, and now she had respiratory failure and had to be put on mechanical ventilation. Pneumonia and emphysema followed, necessitating a chest tube. After five weeks of painful treatment, she died of a pulmonary hemorrhage.

For a 16-month-old girl from rural North Dakota, death came sooner—within two hours after she was admitted to a hospital in January 1998 with a fever of 105.2, seizures, and a rash all over her body. She, too, was treated with a cephalosporin antibiotic. A postmortem culture of her blood and spinal fluid grew MRSA, which had also infected her brain, heart, liver, kidneys, and lungs. She hadn't been in a hospital in the previous year, and neither had any members of her family.

A year later, in January 1999, it happened again: a 13-year-old girl from rural Minnesota was admitted to a local hospital with fever, hemoptysis (coughing up blood), and respiratory distress. On X ray, her lungs appeared infected. Within five hours after admission, she was put on vancomycin and cefotaxime, but neither antibiotic did any good. On the seventh day after her admission, she was dead from cerebral edema and multiple organ failure. MRSA was again the culprit, although she had no recent hospitalizations and none of her family members were either health care workers or had a history of injection drug use.

A month later, in a rural area of North Dakota, a 12-month-old boy was admitted to a hospital with bronchiolitis, vomiting, and dehydration. A chest X ray showed a lung infection. Despite the insertion of a chest tube and treatment with vancomycin and cefuroxime, he was dead within a day, and MRSA was cultured from his pleural fluid and postmortem blood. His sister had been treated successfully for an MRSA buttock infection three weeks earlier, and he apparently got the virulent germ from her.

These cases varied in location—rural and urban—and in race. The 7-year-old was black, the 16-month-old was Native American, and the

13-year-old girl and the 12-month-old boy were both white. Although the extent of community-acquired MRSA is unknown, it is becoming clear that it is now a widespread problem and is no longer limited to health care settings. The CDC now advises doctors to consider MRSA infection as the possible cause in pediatric pneumonia or sepsis in areas where community MRSA infections have been reported. This means treating such patients with vancomycin (in addition to a third-generation cephalosporin) even before the culture results are known.

Another germ that can be deadly and that plagues day care centers is *S. pneumoniae*. Sometimes called pneumococcus, this organism has been for decades the most common bacterial cause of community-acquired pneumonia. Recently it has developed resistance to most antibiotics used for treating it. Children in day care are among the groups most likely to harbor the antibiotic-resistant forms of this microbe. Usually, the beta lactam antibiotics (penicillin and cephalosporin), to which the germ is resistant in lower doses, can be used in higher doses successfully. But other classes of antibiotics cannot be used in such high doses, and the only other class of antibiotics useful against these strains is the quinolones, a new class of antibiotics that must be used very carefully to prevent even more resistant strains from developing. It is not approved for children under 18 except in serious or life-threatening infections.

Children and Antibiotics

Children in day care probably get more antibiotics prescribed for them than children who stay at home. A May 2001 Danish study followed children from birth in 1997 to the age of two years and found that enrollment in a day care center doubled a child's likelihood of being prescribed an antibiotic. Whether this is true in U.S. day care centers is not currently known. But at least in Denmark, the more time parents spend at home with their children (parents in Denmark get six months of parental leave), the less likely they are to have antibiotics prescribed for their kids in the first two years of life. The authors of the study were unable to conclude whether the increased use of antibiotics was the result of actual increases in illness, or merely from parental demands for antibiotic treatment. They did discover that children with older siblings were less likely to get antibiotics, but again they couldn't figure out whether this was due to the increased immunity likely in younger

siblings or to the parents' tolerance of illness without insisting on antibiotic treatment.

One of the most common ailments of childhood, and perhaps one of the most common reasons for the unnecessary prescription of antibiotics, is ear infections—serous otitis media. (Serous otitis media is distinguished from the otitis media that involves bacteria in the middle ear without the presence of fluid.) This is an ailment in which kids get fluid in their ears, usually for much the same reasons they get runny noses. Ears make fluid in the same way the nose does—it just doesn't fall out as easily as it does out of the nose. It can take up to a month to dissipate. So what are you supposed to do about it? In most cases, nothing except wait for it to go away, but parents often insist on getting antibiotics for it, and doctors give in when they shouldn't. The only time when antibiotics might be necessary for otitis media is when the fluid is still present after several months—it's best to see a doctor a few months after an infection to see if that has happened. But usually it doesn't happen. Otitis media almost always gets better by itself.

Infant Botulism

Botulism, which we discussed in chapter 2 and is extremely rare in adults, is somewhat more common in infants (there are about 80 or so cases a year, while 2 or 3 cases of adult botulism per year would be about average). The effects of the disease in infants can range from fairly mild to death—in fact, it is one of the causes suspected in sudden infant death syndrome. Infant botulism probably happens because the kids have not yet developed the full range of intestinal bacterial fauna that in adults helps to fight off the effects of the bacterium. While contaminated honey has been identified as one source of the spores, there are many cases in which no foodborne source can be found. The disease usually occurs in infants under six months old, and it is therefore recommended that kids under one year not be fed honey.

Resistant Acne

In May 2001, a Swedish researcher named Carl Erik Nord made news by announcing at the annual meeting of the American Society for

Microbiology that the bacteria that causes severe acne—*Propionibacterium acnes*—were developing resistance to the antibiotics used to treat the affliction. This may have been news to the reporter who wrote the story, but it shouldn't have been news to anyone else at the meeting: the bacteria that cause acne have been developing resistance to antibiotics at least since the late 1970s. An article published in an Australian medical journal in 1998 summed up the problem. The overall incidence of *P. acnes* antibiotic resistance had increased from 20 percent in 1978 to 62 percent in 1996. The amount of resistance depends on the antibiotic, but the most common resistances are to erythromycin and clindamycin, tetracycline and doxycycline, and trimethoprim. And the continuing use of antibiotics is producing more and more resistant strains. Using long-term antibiotics for acne is probably an outdated treatment whose result will be not to cure acne but to create more resistant strains of bacteria. A vaccine against these species of bacteria is therefore the most likely to produce the desired results, but no such vaccine exists at this time.

Everybody Wants the Medicine

Alliance, Ohio, is a town of about 25,000 people southeast of Akron. It is famous, if we use the word loosely, as the site of a yearly carnation festival every August and the home of the particular variety of carnation that President William McKinley favored as a lapel decoration. But one day in June 2001, the town became famous for something else entirely: bacterial meningitis. Three high school students came down with the disease and two of them died, frightening the entire town, and with good reason. Bacterial meningitis has a fatality rate of 10 to 15 percent, and certain forms are contagious. This was definitely an outbreak, not coincidental individual cases—the CDC performed a polymerase chain reaction test on the *Neisseria meningitidis* type C they found in the three kids, and the bacteria were genetically identical. *Neisseria* meningitis is one of the contagious forms.

Meningitis means a swelling of the meninges, the membranes that cover the brain and spinal cord. The term itself says nothing about what causes the illness. When we call an illness "bacterial meningitis" we are saying what the cause is, but only very vaguely—it's caused by bacteria.

It is indubitably true that a person with bacterial meningitis is infected with bacteria, but what germs these are and why they are making him sick is another question entirely. Many species of bacteria can cause meningitis, although three of them account for about 80 percent of all cases: *N. meningitidis* (meningococcus), *H. influenzae* type b (Hib), and *S. pneumoniae* (pneumococcus).

Meningococcus can be found in the nose and throat of about 5 percent of the population, and it is easily spread by respiratory droplets and close contact with an infected person. But very few people actually get sick from the germ, and no one knows why. Meningococcus most often causes meningitis in children under one year old. It can also occur in military barracks, college dormitories, and other such closed populations. Hib also causes meningitis, but mostly in children who are not vaccinated against it, and not in adults unless they have some predisposing factor, such as being immunocompromised or having had a head injury. Among adults, the most common cause of meningitis is pneumococcus. Children under a year old are the most common victims of the disease, and the cause can be not only Hib and meningococcus, but also *E. coli* or streptococcus.

Acute bacterial meningitis usually begins with a sore throat and a fever, headache and stiff joints, and then vomiting. Most people get better with antibiotic treatment, but a small number become irritable, confused, and drowsy, fall into a coma, and die. And all this can happen within 24 hours in adults, and even faster in children. In infants, the symptoms are less predictable, but often include swelling of the fontanel (the "soft spot" on an infant's skull), fever, vomiting, and a high-pitched cry. This makes early diagnosis essential, and any unexplained fever in a young child is cause for a lumbar puncture to examine the spinal fluid if the infant is increasingly irritable or has vomiting, lethargy, or seizures.

Most kids get their first Hib vaccine at two months. This has made the disease very rare in the United States, and has virtually eliminated it as a cause of meningitis in kids over that age. It still causes considerable trouble, however, for infants under two months old. Most of the meningitis in older children and young adults in the United States is now caused by *N. meningitidis*.

The best treatment depends on the age and health status of the patient and on what organism is causing the illness, but the first-line treatment for adults when the exact organism is unknown is a combi-

nation of antibiotics including penicillin or ampicillin along with cefo-taxime, ceftriaxone, or vancomycin.[*]

Sometimes meningitis appears in sporadic cases here and there; other times there is an outbreak of illness, as there was in Ohio. These outbreaks used to be rare, but since 1991 their frequency has increased. Between July 1994 and July 1997, there were 42 outbreaks of meningo-coccal disease, including four at colleges. Still, outbreaks account for a small minority of cases—less than 3 percent. You're more likely to get the illness alone than in company with others. The outbreaks that do occur typically happen where large numbers of young adults live in crowded conditions.

Military recruits are now routinely vaccinated against meningo-coccal disease (as are students entering the military academies), so rates there are quite low, and the outbreaks that used to occur before the use of the vaccine have stopped. College dorms, though, are another story. From September 1998 through August 1999, there were 90 cases of meningococcal disease in college students reported to the CDC—about 3 percent of the 3,000 cases reported every year. All but three of these cases were undergraduates, and 40 of the 90, or 44 percent, were freshmen. To put things in perspective, there are 2.7 million freshmen who enter college every year, so this is not a common illness. But it can be a deadly one. Five of these 90 students died.

So why don't all students going to college just get the vaccination? Vaccinating 2.7 million kids on their way to college might be a little tricky to arrange, so let's narrow it down to the group most at risk, col-lege freshmen who live in dormitories. That's about 591,000 kids. Now let's consider what would be accomplished by vaccinating this group—or as much of this group as we get to come in and get the shot. There is no doubt that vaccinating all these freshmen would prevent cases of meningococcal infection and deaths from that illness. But meningo-coccal meningitis isn't that common an illness, so you're not going to

[*]In 1942, the first person in the United States to be treated with penicillin, Mrs. Ogden Miller, the 33-year-old wife of the Yale director of athletics, suffered from hemolytic streptococcus septicemia and was cured with about 40,000 units of penicillin—but that was when penicillin was very hard to produce and streptococ-cus would succumb to small doses. Treating such an infection today requires a dose of 1.2 million units. Mrs. Miller didn't have meningitis—treating meningo-coccal meningitis requires 24 million units of penicillin a day for several days.

prevent many cases—you'll prevent between 15 and 30 cases per year in this group and 1 to 3 deaths. But when you look at the cost of all the vaccine (more than $38 million for 591,000 doses plus the cost of administering it and the cost of treating cases of side effects) and compare that number to the benefits of preventing premature death based on lifetime productivity, and the costs of treating the disease itself and its sequelae, which would be lowered or eliminated with the vaccine, you get a very discouraging result: the cost of preventing a single case of meningococcal disease would be somewhere between $600,000 and $1.8 million, and the cost of preventing a single death would be between $7 million and $20 million. The more people you vaccinate, the more expensive preventing disease and death becomes. If you decide you're going to vaccinate all freshmen and manage to get the vaccine to 2.3 million of them, you would prevent 37 to 69 cases at a cost of $1.4 million to $2.9 million per case, and two to four deaths for something between $22 million and $48 million each. If you or someone you know are one of the deaths prevented, of course, you'll have a certain view of the economics. But the people who decide which public health measures should be undertaken aren't considering just you— they have to consider all of us, including people who get sick from all other diseases and those who don't get sick at all. From their perspective, spending $48 million to prevent one death from meningococcal meningitis is not a good use of public health funds. These calculations may seem rather coldhearted, but without considering such numbers no public health program can be rationally undertaken or have any hope of success.

Yet the vaccine is available, safe, and effective, so the Advisory Committee on Immunization Practices (ACIP) makes these recommendations: Freshmen and their parents should be informed about meningococcal disease and of the availability of a vaccine, and should get it if they ask for it. Other students should be similarly informed, and the vaccine should be available to those who want it. The ACIP does not recommend any changes in living situations for freshmen students because the slight increased risk for these students does not warrant it. Still, it's important for students to seek medical attention for unexplained fever, headache, or neck pain.

In Alliance, Ohio ("Home of the Scarlet Carnation"), state and federal health officials did their calculations and decided to vaccinate only the 5,800 students and faculty members in six high schools in the region

after it became clear that this was the population at highest risk. Needless to say, this did not reassure the 30,000 other people in the area, many of whom thought they'd like to be vaccinated, too—and damn the expense. So health officials reminded everyone that they could get the vaccine, if they wanted it, with a doctor's prescription. Of course, this meant that doctors in the area would have to write 30,000 prescriptions—and they'd have to find 30,000 doses of the vaccine. The only vaccine available is called Menomune, and it is manufactured by only one company, Aventis Pasteur. This was not a plan that would prove either easy to implement or universally satisfactory. The vaccine costs $65 per dose, and predictably, Aventis couldn't produce it in sufficient quantities fast enough. The city had to create a waiting list. If the CDC's recommendations had been followed, probably only two high schools would have received the vaccine—the local officials decided to extend it to six schools, but this clearly wasn't enough to make everyone happy. Parents in the school districts bordering the ones that got the vaccinations were incensed, pointing out that their children had plenty of contact with kids from the vaccinated districts—some of them lived right across the street, said one mother. Why should

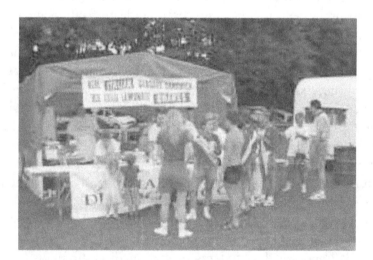

Figure 6 The carnation queen and her court check out the food at the annual Carnation Festival in Alliance, Ohio. Fortunately, the excitement about the meningitis outbreak had died down by the time the festival began. *Alliance Ohio Chamber of Commerce.*

my kid be on a waiting list and have to pay $65 while the kid across the street gets his vaccine free, with no waiting? It is not known whether anyone reminded this mother that if the kid across the street got the vaccine, then that kid wouldn't get the disease and therefore couldn't transmit it to her kid, nor is it known if this information would have consoled her.

That we have a vaccine for a disease is not a good enough reason for everyone to take it, as the example of meningococcal meningitis demonstrates. As for the mayor of Alliance, Ohio, he was just happy that no one else got sick, and that everything had returned to normal before the start of the big carnation festival in August.

6

What Love's Got to Do with It

MICROBES AND YOUR SEX LIFE

You may think that sexually transmitted infections are someone else's problem, but in fact 55 million Americans have at least one sexually transmitted disease (STD). More than 45 million Americans have HSV-2, the sexually transmitted form of the herpes simplex virus, and a million more get it every year. There's no cure for it. Two million Americans have chlamydia infections, 20 million have human papillomavirus (genital warts), another half million have hepatitis B. All of them are STDs. Two old favorites, syphilis and gonorrhea, have returned to haunt us. There have been no recent national prevalence surveys (how many people have the illness) for these two, but gonorrhea incidence (the number of new cases reported) increased almost 9 percent from 1997 to 1998 after a twenty-year decline, and urban areas across the country reported an increase in syphilis in 2000–2001. There are an estimated 150 million new cases of one STD or another every year. In other words, a large minority of Americans are infected, and more are becoming infected every year. No, this is not someone else's problem. It's yours.

Since these diseases are transmitted through sexual contact, taking medicine won't work to stop their spread, even diseases for which we have effective drugs. Behavioral changes and access to affordable health care are also required if we want to eradicate any of these illnesses. Developing new drugs to treat infections is a complex and difficult process. Changing people's behavior, especially in matters having to do with sex, sometimes seems impossible.

STDs are more common in women than in men, and women suffer more from their complications. One of the most serious complications is pelvic inflammatory disease (PID), which can lead to chronic pelvic pain, infertility, and potentially fatal ectopic (tubal) pregnancy. There are various organisms that cause PID, but most cases are associated with gonorrhea and chlamydia.

Women can also pass on STDs to their babies. Syphilis, for example, affects the developing fetus. Other STDs can be transmitted to a child during birth. Gonorrhea, chlamydia, herpes, and human papillomavirus fit into this group. The human immunodeficiency virus* (HIV), which causes AIDS, can infect a child during birth and even through breast-feeding. Babies born with syphilis or herpes can be treated, but the treatments are often only partially successful, and the diseases can lead to severe neurological damage, eye disease, pneumonia, and death.

Teenagers and young adults are at high risk for acquiring STDs. Young women in particular are biologically more susceptible to these diseases, and teenagers in general are more likely to have multiple sex partners and engage in unprotected sex. Young women commonly choose sex partners older than themselves, and this also is risky: gonorrhea rates, for example are highest in females between 15 and 19, and in males between 20 and 24, no matter what their race or ethnicity. The prevalence of herpes infection, also common among teenagers, increases with age. An estimated 5 percent of young men and 10 percent of young women are infected, and anyone can get it—the virus doesn't discriminate by race, economic status, or geographical location. Human papillomavirus is epidemic in women under 25: some studies have found rates between 28 and 46 percent.

Syphilis and gonorrhea, unlike herpes, have an uneven geographical distribution, and rates are highest in southern states, possibly because of poverty or limited access to health care and birth control services. Maryland, Tennessee, North Carolina, South Carolina, Louisiana, Georgia, Arkansas, and Alabama have rates of gonorrhea and syphilis higher than elsewhere in the country. But both diseases are on the rise all over the United States.

*You can, of course, contract HIV without having sex, and in the United States about a third of cases can be attributed to injection drug use (and to a much smaller extent maternal-fetal transmission). Worldwide, however, HIV is overwhelmingly a sexually transmitted illness.

The Versatile Virus

In July 1989, Minnesota held a three-week statewide wrestling camp and 175 high school wrestlers aged 14 to 18 came to participate. Everything went fine until the final week, when many of them started breaking out in rashes around their heads and necks or on their extremities. Several had eye infections. Some had swollen lymph glands, fever and chills, a sore throat, or headaches. The last two days of wrestling were canceled, and within a week after camp ended, 60 kids—35 percent of the wrestlers—were suffering from one or more of these symptoms. Health officials held a clinic at the camp to figure out what the problem was, and when they took viral cultures and examined skin lesions, they had their answer: a virus called herpes simplex I. This form of herpes—transmitted by skin contact with viral lesions of wrestlers or rugby players and not by sexual contact—has its own name: herpes gladiatorum, or wrestlers' herpes.

Herpes viruses are very versatile. They can cause lesions on the mouth, lips, and gums; sores in and around the anus; infections of the finger; eye infections that are a leading cause of blindness; viral encephalitis that affects the central nervous system; and infections of the esophagus, lung, and liver. Without treatment, newborn babies with herpes infection die 65% of the time, and few of the ones who survive will develop normally. There are about 90 or so species of herpes virus that infect all sorts of animals and plants, but the main ones that infect humans are herpes simplex I and herpes simplex II, or HSV-1 and HSV-2. By the time they are 50 years old, 90 percent of Americans have been infected with HSV-1 and have the antibodies to prove it. Although it can be transmitted by contact during sexual relations, it isn't normally considered a sexually transmitted disease. HSV-2 is the one that is an STD, usually transmitted only during sex.

Some viruses, like the ones that cause flu, live best in cold weather. Others, like the polio virus, flourish in warm weather.* Herpes thrives all year round, all over the world. The usual way HSV-2 is transmitted is

*The reasons for this are complex and cannot be reduced to "some like it hot and some like it cold." At least one researcher theorizes that it is caused at least in part by seasonal variations in the amount of sunshine and by the body's varying production of melatonin. There are many light-driven changes in mammalian physiology, and there may be such effects in humans as well. More on this in chapter 8.

by contact with the suppurating lesions on another infected person. The virus is also found in saliva and in secretions from the genital tract of both men and women and unlike many diseases is more easily transmitted when there are active symptoms than when there are none.

Both HSV-1 and HSV-2 can cause genital and facial lesions. Kissing is fun, and often the first place HSV-1 manifests itself is in sores that appear around the mouth. The inside of the cheeks, the lips, the gums, and the skin around the mouth may break out in lesions, which can ulcerate and be very painful. The virus is easily spread from active lesions, but you can be infectious even if you don't have any sores, because the virus is spread in saliva. Although HSV-1 is not considered a sexually transmitted disease, it can be spread in genital secretions as well. Sometimes the eruptions are associated with fever and swollen lymph glands in the neck. After the first infection, the symptoms usually disappear, but the virus remains living inside you—no one knows exactly where or in what form—and is often reactivated, for reasons that are not clear. Exposure to sunlight and injuries to the skin seem to have something to do with it, but these are statistical associations, not definitively proven causes. When HSV-2 is reactivated, the lesions it causes are almost always on the genitals and anus, and HSV-2 is much more likely to reactivate than HSV-1. HSV-1 reactivations, when they occur, are much more likely to affect only the lips and mouth area. These are commonly called cold sores. The symptoms of both viruses almost always remain localized to the site of the initial inoculation of the germ, but in immunocompromised people the virus can spread around the body. In otherwise healthy people, the lesions usually heal within a couple of weeks.

Just because you don't have sores or any other symptoms of HSV-2 doesn't mean you can't transmit the disease. You certainly can. In fact, even though it is more easily transmitted when there are active symptoms, most herpes is transmitted by people who don't know they have it, either because they have no symptoms or because the symptoms they do have are so mild that they are never recognized for what they are—sometimes the lesions are so minor that they can be mistaken for a rash or mosquito bites.

HSV-2 is one of the most common sexually transmitted diseases in the United States. More than 1 million new cases appear every year. Ninety percent of people who are infected don't know they are, although they can transmit it to sexual partners. It is not pleasant. It involves recurrent painful ulcers around the genital area, and although the disorder can

be treated to reduce the number and severity of recurrences, there is no cure. Once you have it, you have it forever. The virus hides out in a latent form in the sacral sensory nerve ganglia (in the lower part of the spinal cord) and reactivates from time to time, infecting the skin.

Twenty percent of the population between 14 and 49 years old is infected with HSV-2, and it is more common in women than in men, probably because male-to-female transmission is more efficient than the other way around. The disease is everywhere—there are no geographical areas where it is more or less prevalent. It is increasing fastest in young people: herpes prevalence among white teenagers in the 1990s was five times what it was in the 1970s. In white adults in their twenties, prevalence doubled during the same time period. Among Blacks, the statistics are even more depressing: seroprevalence—the number of people in a given population who have the virus in their blood—is about 45 percent among African Americans over 12 years old, and about 17 percent among whites over 12.

How can you tell if you have it, and what can you do about it if you do? The signs and symptoms of the infection vary so greatly that the only sure way is a laboratory test of a sample from one of the sores for the presence of the antibodies. There is no treatment that can cure herpes, but there are some antiviral medications that can reduce the severity of symptoms—you've seen them advertised in those self-consciously decorous commercials on TV. Treatment for the first episode is an antiviral medicine: acyclovir, famciclovir, or valacyclovir. You have to take these medicines several times a day for five to 10 days or more to help heal lesions. Episodic treatment is helpful when there are recurrences, and the medicines appear to be useful over many years—acyclovir has been used daily to prevent recurrences for as long as six years, and valacyclovir and famciclovir daily for one year without the emergence of resistance in people with normal immune function. These medicines reduce viral shedding, but it isn't known whether or to what extent they actually prevent transmission of the disease.

The best way to prevent transmission of HSV-2 is to use latex condoms, but they are not 100 percent effective because there may be lesions that a condom doesn't cover, and the virus is easily passed by contact with an open lesion. You should avoid sex completely when symptoms are present and always use condoms between outbreaks. The outbreak of HSV-1 at the Minnesota wrestling camp would probably have been minimized by excluding wrestlers with suspicious skin lesions.

Most mothers of infants who have herpes at birth have no history of clinically evident herpes infection. The virus is not transmitted while the fetus is in the uterus, but during the birth itself as the baby passes through the birth canal. Herpes-infected births are much more common among women who contract herpes late in pregnancy than those who get it earlier—those women who get it early in pregnancy usually pass along antibodies to their babies, who are then immune. The treatment of HSV-infected infants is complex, and must be managed by a physician expert in this area of medicine.

Infection with HSV-2 has one other extremely unfortunate side effect: it makes you as much as five times more susceptible to HIV infection.

The "Foreign" STD

The Italians call it the French disease; the French call it the English disease. Russians blamed the Poles, the Poles blamed the Germans. At least since the sixteenth century, Europeans have been blaming each other—if it isn't English, French, Polish, Russian, or German, then it's Venetian or Neapolitan. Some believe Columbus brought it back to Europe from the New World. No one really knows definitively where it came from, but everyone agrees on this: syphilis certainly didn't come from *us*.* Some believe that syphilis has existed in Europe since antiquity, but that it had always been mistaken for leprosy or some other disease. But confusion reigns, because the best definitive descriptions of the disease in Europe date only from about 1500, and there is no way to know whether this is because it was a newly imported disease or because medical knowledge had progressed to a point where a clear description of the illness was possible. It is probably best to leave aside the question of who should take "responsibility" for the existence of a bacterium, but the question of the origin of syphilis, as of any disease, is nevertheless of some interest.

Even the latest theory, supported by evidence from the analysis of skeletal remains, hasn't settled the question. Two paleopathologists,

*Before expressing contempt for this antiquated idea, think about the world's reaction to the virus that causes AIDS. Whom have we sought to blame for HIV, and why?

Bruce and Christine Rothschild, have concluded that evidence from bones proves a New World origin. They examined several hundred human skeletons from sites in the United States and Ecuador and claim to have shown that southern skeletons (from New Mexico, Florida, and Ecuador) show evidence of syphilis infection, and that the disease has existed in the New World for at least 800 and possibly as long as 1,600 years. That would mean that someone who had visited the western hemisphere may have brought it back to Europe with him. Yaws is a disease caused by a closely related bacterium, and some now suggest that the *Treponema pallidum* that causes syphilis is a genetic mutation of the yaws bacterium, *T. pallidum* subspecies *pertenue*. But others, citing evidence of syphilis in European skeletons as old as 1,700 years, dispute the Rothschilds' conclusions and still believe that syphilis originated in the Old World, or perhaps in both hemispheres simultaneously, and that Columbus had nothing to do with its presence in Europe, where it had been endemic for centuries.

Whatever its origin, it is clear that there was a significant syphilis epidemic in Europe in the early sixteenth century, and the earliest definitive descriptions of syphilis in Europe date from that era. The infection was probably spread widely by soldiers and their camp followers, and it was known even at that time that the disease was sexually transmitted. By 1530 it had received its name, taken from the shepherd hero of a Latin poem written by an Italian, Girolamo Fracastoro. Fracastoro was a physician expert in contagious disease, a professor of logic, an astronomer, a geographer, a botanist, and a poet (a Renaissance man, both literally and figuratively) who in 1546 published a book entitled *De contagione et contagiosis morbis et curatione* (*On Contagion, Contagious Disease, and Its Treatment*), which assured him a lasting place in the history of epidemiology. But the work for which he is perhaps best remembered is another medical treatise, this one published as a long poem called *Syphilis sive Morbus Gallicus*, which translates into English, predictably, as "Syphilis, or the French Disease."* Syphilis is the name of the shepherd hero of the piece.

*Publishing scientific or medical treatises as poems may seem odd, but it was not an unusual practice during the Renaissance and earlier. Lucretius's *De Rerum Natura*, for example, the classic work of natural history, is a lengthy poem. Fracastoro's poem contains a section on Columbus's voyage to America, but he rejected the theory, which had already taken hold, that syphilis was brought back to Europe from the New World.

Figure 7 Fracastoro, the 16th-century Italian polymath who gave syphilis its name and blamed it on the French. Perhaps one of the first, but certainly not the last, to blame a nasty disease on foreigners. *Jean-Jacques Boissard and Theodor de Bry: Bibliotheca chalcographica, hoc est Vertute et eruditione clarorum Virorum Imagines.*

For centuries, doctors assumed that syphilis and gonorrhea were the same illness, but in 1837 a French physician, Paul Ricord, showed that they are caused by two different organisms. He also distinguished the three different stages of syphilis infection. Rudolph Virchow discovered that the infection could move through the blood to internal organs, and by 1876 cardiovascular syphilis had been clearly documented.

Syphilis epidemics occur periodically in various parts of the world, and blaming others for them has been the standard response. In 1862, Arthur B. Stout, a doctor prominent in the American Medical Association, published a fuzzy-brained report called "Chinese Immigration and the Physiological Causes of the Decay of a Nation," in which, with no scientific evidence whatsoever, he accused Chinese immigrants of bringing syphilis into the United States. Such reports were quickly taken up by nativists in Washington eager to codify their bigotry by imposing strict restrictions on Chinese immigration. Later, Italian immigrants would take the rap for *T. pallidum*. Jews and Irish Catholics—no immigrant group is left out when it comes to this kind of bigotry—were usually blamed, respectively, for tuberculosis and cholera in the early part of the twentieth century.

In 1994, there were 2,435 cases of congenital syphilis reported in the United States, the height of an increase that had begun in the mid-1980s. By 2000, the number of cases had declined to 529. No one is quite sure why this decline happened, but it may have been connected to publicity for AIDS prevention encouraging safe sex, which had the effect of reducing other STDs as well, particularly among men who have sex with men, which was the population most at risk for the illness in the 1980s and 1990s. In any case, in 1999 syphilis rates increased for the first time in 10 years to 130 cases. There were 117 cases reported in 2000. In the first six months of 2001, there had already been 115 cases reported. Some have suggested that this may also be connected to the (irrationally) reduced worry about AIDS transmission and the corresponding inattention to safe sex practices among populations at risk. These increases are troubling. Syphilis and AIDS tend to travel together, and people with syphilis have an increased likelihood of acquiring and transmitting HIV because of the genital ulcers that break the skin. Many urban areas, including Los Angeles, San Francisco, Seattle, and Philadelphia, have seen an increase in syphilis among men who have sex with men, and the disease has persistently high rates in rural areas, particularly in the Southeast.

Syphilis is caused by a spirochete, the spiral-shaped bacterium *T. pallidum*, and it is transmitted during vaginal, oral, or anal sex, and more rarely by kissing or close skin contact. Pregnant women can pass it on to their newborn babies. You can't get it from toilet seats, doorknobs, swimming pools, or hot tubs, or from sharing eating utensils or clothing.

An average of about three weeks after infection a sore called a chancre appears in the genital area or wherever syphilis first entered

Figure 8 *Trepona pallidum*, the spirochete that causes syphilis. This is a picture of a rabbit testes experimentally infected with the bacterium. The bacterium is called a spirochete because of its corkscrew shape. *CDC. Photograph by Dr. Edwin P. Ewing, Jr.*

the body. In its earliest stages, it isn't possible to diagnose syphilis by a blood test. You have to actually see the spirochetes under a microscope in a sample taken from one of these sores. Figure 8 shows what the spirochete looks like using a technique called dark-field microscopy. The chancre is usually small and round, and it doesn't cause any pain. Then it heals on its own within a month or two. But this is not the end of the story. If you don't get treatment during this period, the disease goes into its secondary stage. This starts with a skin rash that appears either as the chancre is going away or up to several weeks after it disappears. The rash, red or brown, can appear anywhere on the body and has a particular predilection for the palms of the hands and the soles of the feet. It doesn't hurt or itch. You can also get fever and swollen lymph nodes during this stage, and a sore throat, headaches, hair loss, tiredness, or muscle aches. You are infectious when there are symptoms, either in the first or second stages. This rash clears up on its own, even if you don't get any treatment, but there is still more to come.

Syphilis doesn't go away just because the secondary stage symptoms have disappeared. Instead, it lurks in the body during what is called the latent, or hidden, stage of the disease. The virus lives on and antibody tests for it are positive, although there are no active symptoms. It can still be treated at this point with penicillin, but if it isn't treated, about 50 to 70 percent of infected people progress to tertiary syphilis. This is the stage that causes death due to complications in the skin, bones, central nervous system, heart, and blood vessels. The lesions recur at this stage and can be present for vastly varying periods of time—from as little as two years to as many as 40 years from the onset of the infection. When the lesions occur on the heart, brain, or liver, they can destroy the functioning of those organs and cause death.

Neurosyphilis—that is, disease that affects the central nervous system, causing eye or hearing problems, cranial nerve palsies, or symptoms of meningitis (brain inflammation)—can occur at any stage. It is treated with penicillin and requires an examination of the spinal fluid every six months to look for evidence of disease activity. If there are antibodies still present after two years, then retreatment has to be considered.

If a pregnant woman has syphilis, her fetus can be infected in the womb, and the baby can be born with syphilis. Most infected babies have no clinical evidence of disease at birth, but soon develop a runny nose and a widespread rash. Treatment involves consideration of the mother's health status and what treatment she received during pregnancy, it requires careful follow-up and serologic testing, and it depends on careful diagnosis of the infant.

Gonorrhea

In 1998, 355,642 cases of gonorrhea were reported to the CDC. This was an 8.9 percent increase from the 1997 rate and the first increase since 1985. In 1999, the latest year for which complete figures are available, 360,076 cases of gonorrhea were reported, a 9.2 percent increase over the 1997 rate and a 1.2 percent increase over the 1998 rate. Since reported cases are estimated to account for about half of all annual infections, this means that approximately 700,000 people in America per year get the disease. The increases are particularly acute among men who have sex with men and among teenagers. Among adolescents

during the period 1997 to 1999 reported cases increased 13 percent. Reported gonorrhea rates in the United States are higher than in any other industrialized country—about eight times the Canadian rate, for example, and nearly 50 times the rate in Sweden.

Albert Neisser, after whom the bacterium *Neisseria gonorrheae* is named, was a nineteenth-century German dermatologist, and probably not a very nice man. Armauer Hansen, the Norwegian microbiologist who discovered the bacterium that causes leprosy, generously shared information with Neisser, and Neisser paid him back by trying to claim credit for Hansen's discovery. That leprosy is now known as Hansen's disease is proof of his failure. (Not that Hansen himself was a candidate for sainthood. Having failed to grow the bacterium in a culture dish or any experimental animal, he tried to inoculate leprosy into the eye of a woman without her consent or knowledge. In 1880 he was sued in a court in Bergen, Norway, and found liable for damages to the woman's eye.)

Questions of ethics aside, Neisser was certainly a talented microbiologist, and did legitimately succeed in 1879 in isolating the bacterium that bears his name. *N. gonorrheae* is an aerobic, gram-negative bacterium that lives in the mucous membranes of many animals, including humans. It thrives in a moist environment at human body temperature. There are numerous species of *Neisseria*, but the most important disease-causing forms are *N. gonorrheae* and *N. meningitidis* (which causes meningitis—see chapter 5).

You can be infected with gonorrhea for many years without showing any symptoms, but when the symptoms occur, they are very nasty. In men, there is a urethral discharge of yellowish-white pus that can result in prostatitis, or abscesses around the urethra. Urination may cause painful burning sensations. Women with gonorrhea may have a yellow or bloody discharge from the vagina, bleeding associated with vaginal intercourse, and sometimes painful burning while urinating. In men, the symptoms usually begin soon after infection (although they can appear as much as a month later), but since 80 percent of women with gonorrhea have no symptoms, they can transmit the disease without even knowing they are infected. Getting one case of gonorrhea, far from making you immune, may actually make complications worse the next time you are infected. In women, these complications can lead to sterilization; in men, swelling of the testicles and penis. Arthritis, skin infections, and other problems are common in both sexes.

Gonorrhea used to be pretty easy to treat—penicillin taken either orally or by a shot in the rear end, or tetracycline, another common and inexpensive antibiotic, did the trick. But during the 1980s, resistance to these two drugs began to appear. The CDC responded by recommending other antibiotics, in particular the fluoroquinolones such as ciprofloxacin, the drug later made famous as a treatment for anthrax infection. The CDC regularly tests samples of gonococcal isolates for resistance to antibiotics, and in 1995 they reported more bad news: in eight patients in Washington and one in Colorado, ciprofloxacin-resistant gonorrhea had been found. Even though the number of reported cases of gonorrhea had been steadily declining, the number of cases that were partially or completely resistant began increasing sharply. In 1999 in Hawaii there was an outbreak of a ciprofloxacin-resistant strain, and in Kansas City a strain with increased resistance to azithromycin, a relatively new antibiotic, was found.

The fluoroquinolone antibiotics are excellent medicines. While they are not cheap, they can be given in a single oral dose, and at least until recently, they were very effective. But now the CDC has had to recommend against their use to treat gonorrhea in Hawaii, where resistance to them is endemic. Resistant strains have been known in Asia for several years, and they are apparently now making their way across the Pacific. Now anyone who has traveled to Asia or Hawaii or has a sexual partner who has traveled in those areas can no longer be treated with fluoroquinolones. There is no need for panic yet—the cephalosporins (ceftriaxone or cefixime), and spectinomycin for those who can't tolerate the cephalosporins, are still effective and now recommended—but there is obviously reason for concern, and these developments make prevention that much more important. All the usual recommendations apply, and you already know what they are: limit the number of sex partners, use latex condoms, don't have sex when your judgment is impaired by alcohol or drugs. Gonorrhea, like syphilis, is most common in teenagers and in men who have sex with men, so these groups in particular have to be made aware of the dangers the illness can present. Although researchers are working on it, there is no topical medicine for or vaccine to prevent gonorrhea.

The Most Common STD

Chlamydia is probably the most common sexually transmitted disease in the United States, even more common than herpes-2. There were

more than 640,000 new cases of chlamydia reported in 2000 and more than 290,000 through the first six months of 2001. Since all diseases are underreported, and since most people with chlamydia are not even aware they are infected, it is estimated that there are as many as 3 million Americans a year infected with this disease. The disease is so common that by age 30, half of sexually active women have evidence of having had the disease at some time during their lives. The culprit here is *Chlamydia trachomatis*, a bacterium transmitted during vaginal, anal, or oral sex. It can also be passed on to a baby by an infected mother during birth.

C. trachomatis is a bit unusual for a bacterium. For one thing, the disease it causes resembles other diseases so closely that it was not even thought to be a separate organism until the late 1950s. Moreover, since it is an intracellular parasite that only infects humans and no other animal, it was at one point assumed to be a virus. (The name comes from the way it occupies a cell. "Chlamys" is a Greek word meaning a cloak draped over the shoulders, and the bacterium "drapes" itself around the cell's nucleus.)

In women, the symptoms of the disease include lower abdominal pain, low back pain, nausea, fever, pain during intercourse, and bleeding between menstrual periods. Men may have a discharge from the penis and a burning sensation when urinating. There may also be a burning feeling around the opening of the penis and pain and swelling of the testicles. But three-quarters of infected women and half of infected men don't have any symptoms at all, and it can only be diagnosed by collecting and testing a specimen from an infected cervix or penis, or by testing a urine sample.

For many people, this means that they only realize they have the disease when they get one of its complications. In men, this can mean urethral infection, which can spread to the epididymis (the tube that carries sperm from the testes to the penis), pain, fever, and possible infertility. If chlamydia infects the cells of a woman's cervix, the infection can spread to the uterus and fallopian tubes. The result of this is pelvic inflammatory disease (PID), causing damage to the fallopian tubes, uterus, and ovaries. This can lead to chronic pain and infertility. Babies born to infected mothers are at risk for eye infections and pneumonia. Sexually active teenage girls and young women are especially at risk for this infection because of the characteristics of the cells that line the cervical canal. Babies are treated with oral erythromycin if their mothers were infected during pregnancy.

Prevention is the best approach. Sexual abstinence is a sure way to prevent the illness. Limiting sexual contact to one uninfected partner and using condoms (correctly and every time) are sensible and effective steps. Treatment is relatively simple—but of course you have to know you have the disease before you can get it cured. If you are sexually active, and particularly if you are young, you should be screened for chlamydia at least once a year. All pregnant women should also be screened. Screening programs are extremely effective—wherever family planning or STD clinics offer screening for chlamydia, rates of the disease decline, and the longer the programs are funded the greater the decreases. Screening programs have largely been directed at women (which may be one reason why reported rates for women are higher than those for men—men's infections may be just as common but not as often diagnosed), but men need to be screened, too. Men have symptoms and seek treatment on their own more often than women, but half of men are asymptomatic. This means that reaching them with screening programs is essential.

A single dose of azithromycin or twice-a-day doxycycline for one week are the usual treatments. You have to tell your sex partners if you are infected because they need to be treated, too.

Not All Warts Are Alike

Warts—the kind that appear on the skin, often of older children—are common and benign. There are various treatments that can get rid of them, but most of them go away by themselves after a year or two. Yet there are many viruses—probably 60 or more—that cause warts, and the warts these viruses cause are not all alike. Some of them have consequences that are not benign.

Human papillomavirus (HPV) is transmitted during sexual contact. This is the virus that causes genital warts, although a person can be infected with HPV and have no symptoms. At any one time, 20 million Americans are infected with HPV, and the disease is quite contagious—there are 5.5 million new infections every year. There are 30 different types of HPV that can infect the genitals. Some cause warts, others don't, and most of these infections clear up by themselves, although reactivation is always possible. There are treatments for the warts, including some that can be done by the patient such as using podofilox

and imiquimod, which are creams or gels that can be applied directly to the warts. A health care professional can also apply the topical preparations podophyllin resin, trichloroacetic acid (TCA), or bichloroacetic acid (BCA). Injecting interferon directly into the warts can also be effective. A doctor can perform cryotherapy (freezing) or surgical removal. But the types of HPV that don't cause warts are more common. When these latter infections are persistent, there is no effective treatment, and they can lead to cervical, penile, and anal cancer. Persistent infection of the cervix with certain of these viruses is the chief risk factor for cervical malignancies—there are four of them that account for 80 percent of cervical cancer. Although researchers are now working on it, there is still no vaccine to prevent HPV.

Sexually transmitted HPV is very common among reproductive-age Americans. Estimates are that 75 percent of that population has been infected at one time or another, and that currently 15 percent of Americans between 15 and 45 have HPV. In a study of female college students, 43 percent became infected during the three-year period of the study. Among women under 25, studies show prevalence rates of between 28 and 46 percent.

Most HPV infections don't have any symptoms, and there isn't any way to tell if you have it by a physical exam or even by an examination of a sample of cells under a microscope. Some types of HPV cause abnormal Pap smear results, but the most reliable way of finding out if you are infected is through a test that reveals the presence of HPV DNA. The most common serious consequence of HPV infection is cervical cancer. Most types of the virus don't lead to cancer, but more than 90 percent of cervical cancers are preceded by some form of HPV infection. This doesn't mean that HPV is the cause of cervical cancer—many other factors are involved, including the number of sex partners, age at first intercourse, other infectious agents besides HPV, cigarette smoking, certain dietary deficiencies, and immunosuppression. But HPV infection is certainly one of the most powerful risk factors, and preventing HPV infection would surely lower the number of cases of cervical cancer. The best method for detecting cervical cancer at its earliest possible stage is an annual Pap smear and pelvic examination for all women who are or who have been sexually active or who have reached age 18. After a woman has had three or more consecutive satisfactory normal annual examinations, the Pap test may be performed less frequently at the discretion of her physician.

HPV infection also puts you at greater risk for anal cancer, a relatively rare malignancy but one with an increasing incidence. Over the past 25 years there has been a substantial increase in anal cancer among men who have sex with men, and a somewhat smaller increase among women. Although anal Pap smears are not widely used for this purpose, they may be as sensitive a detector of anal cancer as they are of cervical cancer, and there is reason to believe that screening for anal cancer among susceptible populations may be worth the effort. There is no test for penile cancer—it can only be found by biopsy of a suspicious lesion found on physical examination.

Prevention of HPV infection is fraught with problems. The spread of HPV is a function of three factors: how easily the virus can be transmitted, the duration of the infection once you have it, and the number of new partners an infected person has. Condoms probably don't benefit either men or women because HPV can involve the external genital skin and not just the mucosal linings of the reproductive tract. There is some evidence that certain microbicides may be effective in inactivating HPV, but this is still uncertain and there is currently none approved for use. There is no pill or shot for HPV as there is for syphilis and gonorrhea. The treatment of genital warts reduces viral DNA and mRNA in subsequent biopsies, but nearby clinically normal skin or mucosa often contain HPV, which may explain why people get reinfected.

What should you know if you are infected? You should know that there is a high prevalence of the infection among sexually active adults, that the infection is contagious through sexual activity, and that it can be persistent for an indefinite amount of time after treatment. You should know that condoms don't help prevent transmission, but that their use for the prevention of other STDs is still important. Because HPV is so common among sexually active people and because the duration of infectiousness is unknown, the value of disclosing a past infection to a sexual partner is unclear. Still, frank discussion of past STDs is almost always a good idea.

A Venereal Parasite

Syphilis, gonorrhea, herpes, papillomavirus, chlamydia—you'd think that would be enough bacteria and viruses to worry about, especially at a moment when you're preoccupied with something more pressing.

But there's more. The cause of trichomoniasis is neither a bacterium nor a virus. It's the cause of a microscopic parasite, a protozoan called *Trichomonis vaginalis* that (despite its name) infects both men and women. Men usually have no symptoms, although they can sometimes have urethritis, an inflammation of the urethra that causes pain inside the penis or slight burning after urination or ejaculation. Women are more frequently symptomatic, and typically have a frothy and malodorous yellow-green discharge and discomfort during intercourse and urination. The genital area can also be irritated or itchy.

The infection is easily treated with a regimen of metronidazole, the only medication available in the United States for the treatment of trichomoniasis. Cure rates are very high—90 to 95 percent—and treating sexual partners probably increases the cure rate even further. This means treating asymptomatic men, who can still transmit the infection if they are not treated. The disease can also be transmitted by vulvar contact between women, so female sexual contacts of infected women must also be tested and treated.

The infection can be diagnosed in women by pelvic examination— small red sores on the vaginal wall or cervix are usually apparent. Tests can also be done on a sample of vaginal fluid or urine. It is harder to detect the parasite in men.

The usual admonitions apply: abstain, use condoms, limit the number of sex partners, limit sexual contact to one uninfected partner. If you have symptoms, get treatment and counsel from a health care professional, and don't have sex again until you're better.

Eradication Is Not an Option

Why can't we eradicate these diseases? We often speak of the possibility and desirability of eradicating diseases, and it does seem that it should be possible to eradicate at least those whose causes we know and for which we have good and effective treatments. This would seem to apply to most of the sexually transmitted diseases we've discussed here. But in all of history, only one disease, smallpox, has ever been eradicated, and that happened only after a vaccine for the illness had been in use for 181 years and even then only after we had designed and carried out an elaborate and expensive program of vaccination that required worldwide cooperation through the United Nations. People sometimes say, either wistfully

or angrily, something like this: "If we could just stop spending money on weapons, we could eliminate . . ." or "if only we had the political will to do so, we could eradicate . . ." and then they fill in the name of some terrible illness. Unfortunately, it isn't that simple. True, a handful of infectious diseases undoubtedly can be eliminated if we devote the money and international effort to doing so. With a few others it may be possible to eliminate some of the worst consequences of infection. But there are some diseases that cannot be eliminated, no matter how earnest our efforts and how fervent our desire to do so, because the nature of the microbes that cause them makes this impossible.

Smallpox is in many ways a perfect candidate for elimination—in fact, among infectious diseases, it is surely the easiest to extinguish. It is relatively noncontagious, usually requiring very close contact to be transmitted.* The disease is easy to diagnose by its characteristic skin lesions or by antigen detection. It only infects humans—you can't catch it from an animal, wild or domestic. Immunity develops after one infection, and there is no carrier state, or period in which a person can be

*There is an oft-repeated account of the British general Lord Jeffrey Amherst using germ warfare against American Indians by giving them smallpox-infected blankets. There is no question that Amherst wanted to do this—there is a letter extant dated July 13, 1763, from Amherst suggesting just such a tactic to Colonel Henry Bouquet. There is no proof, however, that the project was ever carried out. Undoubtedly many Indian tribes suffered terribly from smallpox infection, as did the British troops—there was the technique of variolation, prophylactic infection with smallpox itself, but it was not widely used, and there was no really safe vaccine. Any infected blankets that Amherst had in his germ warfare arsenal must have been infected by British soldiers. But that any of the Indians' suffering was caused by Amherst's plan to use germ warfare is unknown. Even the nature of the smallpox virus is controversial in this respect. *Harrison's Principles of Internal Medicine* (14th edition, 1998), the standard text used in almost all medical schools, states that "smallpox was a relatively noncontagious disease whose transmission required close contact," and that this is one of the characteristics of the virus that made it easier to eradicate. The Centers for Disease Control says that "smallpox is spread from one person to another by infected saliva droplets that expose a susceptible person having face-to-face contact with the ill person," and adds equivocally that "contaminated clothing or bed linen could also spread the virus." As for time of survival on environmental surfaces, the CDC says, "The smallpox virus is fragile and in the event of an aerosol release of smallpox, all viruses will be inactivated or dissipated within 1–2 days." Whether or not the virus can survive on environmental surfaces such as blankets and for how long is unknown, nor is it known how infectious such blankets would have been to the (hypothetical)

infectious but asymptomatic. We have had a vaccine for it for more than 200 years, and have had ample time to perfect that vaccine so that it is virtually 100 percent effective and 100 percent safe. The vaccine is inexpensive and highly resistant to temperature changes, chemicals, and drying, so it can be shipped over long distances without refrigeration or other special requirements. It requires only one dose and can be given to infants at birth. Several decades can pass before a booster vaccine is required. The smallpox virus existed in only two strains, neither of them resistant to the vaccine, probably partly because the illness has never been treated, only prevented. In other words, smallpox was in many ways ripe for elimination. And even then it took us almost 200 years after we knew exactly how to prevent the disease to actually eliminate the germ.

Now think about a microbe that isn't as cooperative as the smallpox virus. Think, for example, of the bacterium that causes tuberculosis. There is a vaccine for TB, rarely used in the United States but sometimes used, mostly on children, in other countries. The problem with it is that it isn't very effective—you can still get TB even if you've had the vaccine. Obviously, for anyone who already has TB, the vaccine is completely useless. So that leaves treatment. Treatment of TB requires repeated doses of medicines over a period of six months or more (isoniazid, rifampin, pyrazinamide, ethambutol, streptomycin—you have to take more than one drug to prevent the emergence of resistant mutations in people in whom the bacterium is rapidly

British soldiers who delivered them as compared to any Indians who received them. Most people who get smallpox recover, and undoubtedly many British soldiers had recovered from the illness and were therefore immune. Some were surely not. The last outbreak of smallpox, in Bangladesh in 1972, killed an estimated 20 percent of those infected [See: Foster S.O., Ward N.A., Joarder A.K., Arnt N., Tarantola D., Rahman M., Hughes K. Smallpox surveillance in Bangladesh: I—Development of surveillance containment strategy. Int J Epidemiol. 1980 Dec;9(4):329–34.] Donald Aislee Henderson, who headed the United Nations effort to eradicate smallpox, says that his group heard many tales of smallpox suddenly breaking out in a place that had housed smallpox victims months or even years earlier, but they were unable to document any such case (personal communication, December 29, 2001). There are, however, reliable accounts of smallpox infection being passed in laundry, which can remain infectious for a day or two or until it is immersed in water. So it is possible, though not likely, that Amherst's biological warfare could have succeeded. We must at the very least give Lord Jeffrey full credit for his bad intentions.

reproducing). If you skip doses, the treatment is not only ineffective, but actually makes the disease worse by creating strains of bacteria resistant to the medicines. You can't trust people to take their medicine—they forget, they misplace it, they just don't feel like it today, they have a million excuses—so you have to actually go out and find them where they live, give them their medicine, and watch them swallow it. This is exactly what health care workers do in the United States, in a program called DOT (directly observed therapy), and of course there aren't enough people to do it, and even when there are enough people to do it they don't always succeed. If the person is actually infectious—that is, if the TB is in the lungs or throat—he has to be isolated from other people while he's treated. And if he's sick and bedridden, he has to be isolated in a hospital room, usually for several weeks—not just an ordinary hospital room, but a specially designed room that requires expensive and elaborate equipment, including a ventilation system that creates negative air pressure inside the room (the bacillus is airborne) and vents it through a series of filters to the outside.

So TB treatment can be very expensive and is only partially successful in the United States, where a highly sophisticated and efficient health care infrastructure is in place. Now imagine what it would mean to establish a TB treatment program in a Third World country where there are hardly any health care workers, no clinics or hospitals worthy of the name, severely limited communication and transportation facilities, a population widely dispersed in rural areas, rampant HIV infection that makes TB infection much more common and severe, and where, on top of all that, there's a civil war going on with half the population trying to *kill* the other half, not *cure* them. And we haven't even discussed the side effects of the medicines, which can include everything from vomiting to hearing loss and at least one of which (isoniazid) can cause fatal liver disease if you drink alcohol while you're taking it.

Even with all of these barriers, it is not completely out of the question that tuberculosis might be eliminated, at least in some places. Infectious TB is relatively easy to diagnose by a simple laboratory test. The illness is treatable and curable, and treatments costing as little as $18 are available for simple cases. While asymptomatic people can carry the disease, they can nevertheless be identified by skin testing, and reactivation TB can often be prevented by administering isoniazid to people found to be at risk. The only other reservoir of the bacterium is

in dairy cattle, and this form is rarely associated with human infection. Directly observed treatment-short course (DOTS) is highly effective, even though it is difficult to administer. Eliminating it globally may present a challenge currently too difficult to overcome, but extinguishing it in a highly developed country is conceivable even despite the staggering barriers we've just described.

Yet there are diseases that are even harder to eliminate than tuberculosis. In fact, there are diseases, quite common ones, that cannot be eradicated no matter what we do, even though they are well understood and vaccine-preventable. Pertussis, or whooping cough, is a good example. Pertussis is caused by a bacterium called *Bordetella pertussis*, which was first isolated almost 100 years ago. It causes paroxysms of coughing, often followed by vomiting, and its complications include hypoxia (lack of oxygen), pneumonia, seizures, encephalopathy (brain disease), and malnutrition. Infants and young children are most at risk, and pertussis is deadly. Worldwide, it kills approximately 400,000 people per year. It is very contagious—when one person in a family gets it, there is a 90 percent likelihood that the rest of the susceptible people in the household will, too. There is a vaccine for pertussis, and a large majority of infants and children up to six years old in the United States get it. But the vaccine has certain serious drawbacks. While it is effective in preventing severe disease, it may not prevent acquisition of the infection, and it doesn't prevent mild disease. Even mild disease is highly contagious. The vaccine requires multiple doses, so the delivery system for it has to be highly organized and requires considerable cooperation from patients and their caregivers. After you've been vaccinated, you're immune only for a limited period of time, and if they don't get boosters regularly, older children, adolescents, and adults will become susceptible—and the vaccine is not licensed for people over the age of six years. Pertussis is difficult to diagnose, especially in the phase in which it is most infectious, so control of outbreaks by quarantine and other measures is difficult or impossible. To prevent transmission, you have to make more than 90 percent of the population immune, which isn't attainable even if you vaccinate 100 percent of them. The vaccine is at best 90 percent effective. In fact, the latest reports from the CDC indicate that pertussis is actually increasing among infants, providing a steady reservoir of infection. Eradication of pertussis is simply not feasible.

Among the STDs we've discussed in this chapter, hepatitis B is realistically a candidate for eradication. There is a very effective vaccine that

can be given at birth, and while most chronically infected people remain infected for life, their level of infectivity declines over time. Humans are the only known host for the virus. If we can sustain elimination of transmission long enough for people with chronic infections to die, we could conceivably make the virus extinct. The key is preventing perinatal and infant disease by universal infant vaccination. Young children, adolescents, and adults can be given catch-up vaccinations. Identifying infants who need this postexposure treatment is not difficult in industrialized countries, although it may present challenges in nations that do not have substantial health care infrastructures.

Congenital syphilis is probably more difficult and expensive to eradicate, but it is within the realm of possibility because it can be prevented if infected pregnant women are treated with penicillin. There are no nonhuman hosts of the syphilis bacterium, tests for the disease are very accurate, and early syphilis can be treated with one injection of an inexpensive medicine—all factors that suggest eradication is possible. Eradication of any disease is easier in industrialized countries with good health care systems than it is in other parts of the world, and this certainly holds for congenital syphilis. Elimination of congenital syphilis will require that syphilis rates be reduced, which the continuing national campaign against the disease has significantly done. Between 1990 and 2000, syphilis rates fell 90 percent. Rates among blacks and Hispanics remain considerably higher than among whites, and work with these groups is urgent. In addition, the CDC is concerned with outbreaks of syphilis among gay men. Nevertheless, there is now renewed hope that sexually transmitted syphilis might also be eliminated in the United States.

As for the other STDs, we are going to have to continue trying to encourage people to behave in ways that prevent transmission, but always with the awareness that eradication of these diseases is little more than a hopeful fantasy, and that we will live with these microbes, for better or worse, for the rest of our lives.

7

Wild Kingdom

PETS AND THEIR GERMS

Some people never learn. In October 1996, a three-week-old boy was brought into a hospital emergency room in Arizona suffering from a fever of 103.6, plus vomiting and bloody diarrhea that had been going on for more than two weeks. (Why it took two weeks for the parents to bring him in is almost certainly another story.) When the staff took stool and blood cultures to try to figure out what was going on they found *salmonella*, serotype IV 44:z4,z23-, an extremely rare type of the bacterium. In trying to figure out where this odd serotype came from, they examined a sample of stool from the family's pet iguana and found exactly the same rare serotype. There was no question that the infant had been infected by the iguana. His family, prudently, gave the iguana to a relative. Then a month later, imprudently, they let the infant spend two days with that same relative. Forty-eight hours later, the kid was treated again at the emergency room for fever and diarrhea, and the same unusual serotype of *salmonella* was found in his stool.

In April 1997 in Kansas, a six-year-old came down with bloody diarrhea, abdominal cramps, vomiting, and a fever of 104.9. *Salmonella* was the cause again—serotype tymphimurium was found in his stool. They gave him antibiotics—ceftriaxone and amoxicillin/clavulanate. He got better. Nine days later, his little brother came down with the same symptoms, and *salmonella* was found in his stool. No one else in the family was ill. The boys had a corn snake that they kept in their bedroom and handled regularly, and—you guessed it—the snake had *sal-*

monella tymphomurium in its stool as well. The parents had no idea that you could get sick from a snake.

The next month, in Massachusetts, another boy, this one an eight-year-old with a congenital immune deficiency, got severe vomiting, abdominal cramps, bloody diarrhea, and headaches. His stool had *salmonella* serotype St. Paul in it. Three days before the boy got sick, his parents had bought two iguanas from a pet store. They didn't know that iguanas could carry *salmonella*, and when they returned the animals to the pet store, they learned that the pet store owner didn't know this either. The iguanas were never tested, but it's easy to guess what they would have found if they'd examined their stool.

In December 1998 a previously healthy five-month-old suddenly died at home in Wisconsin. They did an autopsy but were unable to find any obvious reason why the child had died. But when they did a culture of a heart blood sample, they found *salmonella* serotype Marina and concluded that the cause of death was septicemia caused by this organism. They found the exact same serotype in the stool of the family's pet iguana. The infant had never even come in direct contact with the animal.

Figure 9 An Eastern box turtle, close up. This one is eating a snail. Turtles carry salmonellosis, and you can catch it from them. *Photo by Dorothy DeLisle, California Turtle and Tortoise Club.*

That consumers and pet store owners generally don't know about the dangers of getting *salmonella* from reptiles has prompted the Centers for Disease Control (CDC) to encourage states to establish educational programs, including legally required point-of-sale information to anyone purchasing reptiles as pets. Iguanas, snakes, turtles—all popular pets. But you might want to rethink the idea of getting one (or reconsider your relationship with the one you already have) when you read stories like these and hear what the CDC recommends about pet reptiles:

- People should always wash their hands thoroughly with soap and water after handling reptiles or reptile cages.
- Children under five and people with weak immune systems should avoid contact with reptiles.
- Reptiles should be kept out of the house if there are children under one year old. If you're having a baby, you should remove any reptiles from the house before the baby arrives.
- Reptiles should not be kept in child care centers.
- Reptiles should not be allowed out of their cages in a home.
- Reptiles should be kept out of kitchens or food preparation areas. Kitchen sinks should not be used to bathe them or to wash their cages, dishes, or aquariums. If bathtubs are used for this purpose, they should be cleaned thoroughly and then disinfected with bleach.

Many reptiles are permanently colonized with *salmonella* species and intermittently shed them in their stool. Over the past 20 years there has been a steady upward trend in the number of reptile-associated *salmonella* serotypes isolated from human blood and stool samples. And this has happened despite a 1975 ban on the commercial distribution of pet turtles less than four inches long, a ban that led to a 77 percent decrease in turtle-associated *salmonella* serotypes isolated from humans. The popularity of snakes and iguanas as pets has apparently more than made up the difference. There are an estimated 7.3 million reptiles owned by about 3 percent of households, and since most reptiles won't breed in captivity, the majority of pets are captured in the wild. Iguanas in particular have become extremely popular—almost a million of them are imported into the United States every year. The CDC believes that if educational programs are ineffective, then the sale of other reptiles should be restricted as they are for turtles. There are roughly 93,000 cases of *salmonella* poisoning a year that are attributable

to contact with pet reptiles or amphibians. Most of these cases are infants, young children, and immunocompromised people, and in these high-risk populations the infections can lead to serious complications such as septicemia and meningitis.

Reptiles are not the only problem. In April 1999 the Missouri Department of Health found a cluster of people—40 of them—who had come down with *salmonella* infections, all of them serotype tymphimurium. Ninety-seven percent of them, they discovered, were exposed to chicks, ducklings, or young turkeys. The ones who washed their hands after handling the young birds got sick less than those who didn't.

The next month, 21 people in Michigan also came down with salmonellosis—they were of various ages and both sexes, and the serotype in all of them was the same: infantis. In addition to their symptoms of diarrhea, fever, and vomiting, they also had in common the fact that they had handled chicks, ducklings, or other fowl just before falling ill. By September, the CDC had determined that all of the infected fowl had come from one hatchery that was shipping 100,000 birds a week, mostly for raising in backyards for egg and meat production. Sometimes, especially during the Easter season, people get ducklings or chicks as pets—these animals probably don't make great pets in any case, and keeping them in the house is especially ill-advised if there is a child under five or an immunocompromised person living there.

You're going to read in this chapter about some pretty nasty things you can catch from the pets you love, so we should say before going any further that we like animals. We have pets, and we've had many different ones over the years. And although research in the matter is still in its infancy, there are some studies that suggest that having pets can reduce stress and actually enhance health. But it is our unpleasant duty to inform you that not all of them are as sanitary as we might like them to be. It doesn't end with reptiles and chicks, either. Parrots and parakeets, cats and dogs, ferrets, wild animals in petting zoos—they all can carry diseases and transmit them to humans. But don't give up your pets yet.

Toxoplasmosis

You may have heard that you can get something called toxoplasmosis from your cat. *Toxoplasma gondii*, the organism that causes the disease,

is a protozoan parasite that infects most species of warm-blooded animals. It has a fairly complex life cycle consisting of three different stages. The first is the tachyzoite, the form that causes acute infection by invading cells. The second is called a bradyzoite, and in latent infections this stage lives in tissue cysts. The third stage is the sporozoite, a form that lives inside an oocyst, a kind of egg shell, and can survive in the environment, where it can be picked up by animals and people. Cats are often a link in the chain—they excrete uninfectious oocysts in their feces, where, given enough time and the right conditions, the oocysts sporulate (hatch, in a sense) and become infectious. The parasites like warm, moist soil, where they can survive for up to a year, although both cooling and heating will kill them. You don't need a cat to become infected, although contact with infected cat feces is one route. Birds, sheep, pigs, goats, and horses have also been found to carry *T. gondii*. If you eat inadequately cooked infected meat or something that has come in contact with infected meat, you can get the parasite that way. You can get it from soil in your garden that has the oocysts in it. It's impossible to know what percentage of disease is caused by cats, infected meat, or soil, but there is a statistically significant decrease in risk for infection if you're a vegetarian. Most serious of all, for reasons we will explain below, an infected pregnant woman can transmit the disease to her fetus, which can be born infected.

It is likely that you already have the parasite—an estimated 23 percent of the U.S. population is infected with it—but it rarely causes any symptoms, so most people never have to worry about it. Even when it does cause symptoms, they are usually not very serious (they resemble symptoms of the flu), and they go away by themselves without any treatment. But some people do have to be concerned about coming in contact with this protozoan, in particular pregnant women and people who are immunocompromised either because they have a disease that affects their immune systems or because they are undergoing certain types of chemotherapy. Infants infected before birth can be born retarded or with other severe mental and physical problems. The immunocompromised can suffer damage to the eyes and brain. Infants are often born with no symptoms, but if untreated develop early learning disabilities and, sometimes decades later, visual disabilities. If babies are treated with pyrimethamine and a sulfonamide for a year, 70 percent of them will develop normally. Estimates of the number of infants born with toxoplasmosis vary, but at least 400 a year, and by some calculations as many

as 4,000, are born infected. Screening programs are not commonly used in the United States, but a program instituted in Massachusetts and New Hampshire during the years 1986 to 1992 found 52 cases of congenital toxoplasmosis in 635,000 infants screened. The cost of the screening program compared to the cost of raising visually or mentally impaired children demonstrated that the program was cost-effective as well. The CDC would like to see more such programs established elsewhere.

There are various blood tests for toxoplasmosis, and some are quite sensitive and accurate. They detect not the protozoan itself, but antibodies to it in the blood. One problem with this is that the tests don't distinguish between a current infection and one that has occurred in the past because the antibodies can be detected for as long as 18 months after an infection has entirely cleared up. Usually the disease doesn't require any treatment, but pregnant women who have the illness can be treated with the antibiotic spiramycin or with sulfadiazine or with a combination of sulfadiazine and pyrimethamine. Which drug regimen is used depends on the gestational age of the fetus and whether the fetus itself is already infected. This doesn't completely prevent infection at birth, but it reduces its probability by about 50 percent.

Toxoplasmosis is a serious health problem. Although few otherwise healthy people suffer from its effects, 750 people, most of them very young or not so healthy, die from it every year, and about half of them got the disease by eating contaminated food. This makes toxoplasmosis the third leading cause of food-borne deaths in the United States. Preventing food-borne toxoplasmosis isn't hard if you follow the procedures discussed in chapter 3 for food handling. Cooking beef, lamb, and veal to 145°F, pork, ground meat, and wild game to 160°F, and whole poultry to 180°F in the thigh is enough to make the parasite harmless. Freezing the meat will also destroy *T. gondii*. Washing vegetables and cleaning off the cutting board, as you would anyway (wouldn't you? See chapter 3), will also help. Pregnant women should avoid gardening or wear gloves if they do garden, get someone else to change the cat litter, keep cats indoors where they can't eat infected mice or other prey, and not adopt or handle stray cats. Cats should be fed only commercial canned or dry cat food or well-cooked table food, and never raw or undercooked meat. These are fairly simple rules, and if you follow them, you won't have to think about euthanizing the cat just because you're having a baby.

The Kids Who Played with the Kittens

On August 12 and 13, 1994, three kids from Pompano Beach came into the emergency room of the same hospital in Broward County, Florida, with a sudden onset of seizures, coma, and respiratory depression. Each of them needed a breathing tube to stay alive. All three—they were 5, 6, and 11 years old—had been perfectly healthy kids up to this moment. They had swollen lymph glands, but lab tests showed that they didn't have viral infections, metabolic disorders, or any evidence of having ingested poison. By the end of September of that year, two more kids, a nine-year-old boy and a three-year-old girl, had shown up at the same hospital with the same symptoms. These kids were from Fort Lauderdale and Boynton Beach, not far from Pompano. The kids were all treated with antibiotics and anticonvulsants, and they all got better in a week or two with no apparent aftereffects. But needless to say, the hospital staff were wondering what was going on here, so they started an investigation. They quickly learned that the only thing the kids had in common was that they had all handled or petted stray kittens. None of them reported having been scratched or bitten. They sent some blood samples to the CDC, where they did indirect fluorescent-antibody testing to try to determine what the problem was. They found antibodies to a bacterium called *Bartonella henselae*, the germ that causes cat scratch disease.

Cat scratch disease is a relatively new ailment. The disease was first described only in 1950, and no one had any idea what caused it until 1992 when blood and other microbiologic studies determined that *B. henselae* was the culprit. Cats don't get sick from this germ—they can carry it for years without showing any symptoms—and it can be transmitted among cats by fleas. Fleas by themselves probably don't cause the disease in humans—direct contact with cats is required. Two other species of *Bartonella* have also been identified in cats, *B. clarridgeiae* and *B. koehlerae*, the first of which can also cause cat scratch disease.

Cat scratch disease occurs more commonly in the fall and winter, probably because domestic cats breed in the spring and infected kittens begin to appear in the fall. The cases described above are exceptional— usually cat scratch disease comes and goes by itself without any treatment, but serious cases can involve the brain, the eyes, the heart, and the bones, and these more complicated cases are treated with the antibiotics trimethoprim-sulphamethoxazole, ciprofloxacin, or azithromycin.

How likely is your own cat to be infected? Somewhere between 14 percent and 44 percent of cats, depending on the study, have been found to be carriers of the bacteria. Treating cats with the antibiotic tetracycline helps reduce the infection, but it isn't certain that it permanently eliminates it. Since it isn't entirely clear how the disease gets from cats to humans, it's best to minimize contact with stray animals, to avoid playing roughly with cats in a way that they scratch or bite hard enough to break the skin, and to get routine veterinary care for your cat that will prevent flea infestation. Bite and scratch wounds should be washed and disinfected promptly.

Rabies and Other Dog Problems

Swenson's Wild Midwest Exotic Petting Zoo is apparently a popular attraction in Clermont, Iowa, and in August 1999 they featured a bear cub. It was really cute, and it let people touch it and wrestle with it. Occasionally it would nip at people, but it never really hurt anyone. The cub was also brought to a celebration at a barnwarming party in Holy Cross, Iowa. By August 27, when the cub died after several hours of acute nervous system symptoms, about 400 people had had contact with the animal. This was too bad, because the bear cub died of rabies, a transmissible disease that, if left untreated, is always fatal.* It is impossible to know if anyone contracted the rabies virus from the cub, so the Iowa Department of Public Health began a campaign to try to trace everyone who had had contact with the bear in order to treat them all prophylactically, the only sure way to protect them. Rabies became more common in animals in the United States and Europe during the 1980s and 1990s, and it continues to spread in the wildlife population. While rabies is quite rare in humans in the United States, no one who touched this bear would want to skip the prophylactic treatment.

*Almost nothing in medicine is ever 100 percent anything, but rabies is so rarely survived that calling it 100 percent fatal is close enough for all practical purposes. There are a few people in the world who have survived HIV infection without treatment, but no one would claim on that basis that HIV is therefore not 100 percent fatal. Similarly, there are a handful of people in the world who can apparently survive without sleep, but to claim that therefore sleep is not a universal requirement of human beings would be absurd.

Although wild animals can be and often are rabid, dogs are the most common transmitter of rabies to humans, and there are about 30,000 cases worldwide every year, very few of them in the United States, where routine vaccination of pets has practically eliminated them as source of the virus. In fact, in the United States cats are more likely to be rabid than dogs. Cats, dogs, and ferrets should all be vaccinated against rabies by a veterinarian.

But dogs in this country carry other diseases that, if not as serious as rabies, may be of more immediate interest. Toxocariasis is one of these, and *Toxocaris canis*, a parasitic nematode, is the culprit. This is usually called dog roundworm (cats can have it, too, in the form of *Toxocaris cati*, but it is much less common in cats). *T. canis* usually lives its entire life cycle in dogs, but humans can also be an accidental host of the parasite. Puppies (and sometimes kittens) are infected by their mothers, either in the uterus or after birth through suckling. The nematode, which has been living in the mother's tissues in an undeveloped state, then grows in the puppy's or kitten's intestines and is shed in feces. You get it by oral contact with the *toxocara* eggs, expelled in dog feces into soil or onto other surfaces. Kids, as you might expect, are more likely to get it than adults, but adults, especially dog owners, are also at risk. There are two ways in which the nematode infects people. The first is called visceral larva migrans. The nematodes don't usually reproduce in humans; instead, the larvae migrate through organ tissues, causing inflammation as they go. While most infections are asymptomatic, preschool children often get sick with fever, malaise, loss of appetite, and weight loss. Death from this parasite is rare, but when it occurs it is usually because the heart, the lungs, or the brain have become infected.

When *toxocara* invades the eye, the illness is called ocular larva migrans. A mass develops around the entrapped larva that looks like a retinoblastoma, a malignant tumor. Careful diagnosis is important because a retinoblastoma often requires removal of the eye—something you don't want to do if the problem is a temporary infection caused by a nematode parasite. The infection causes eye pain and inflammation in addition to visual disturbances. Unlike visceral larva migrans, which almost always affects young children, it is usually older children and adults who get the ocular form. Usually toxocara infections of both kinds resolve by themselves without treatment. There are some drugs that can be used, but their effectiveness is questionable. Treating the ocular form usually doesn't work at all.

Dogs also carry *Ancylostoma braziliense*, another nematode parasite, known commonly as hookworm. This one causes an illness whose symptoms are even more revolting than those of *toxocara*. The scientific name of the disease is cutaneous larva migrans, but its common name is much more vividly descriptive: creeping eruption. The larvae hatch from eggs in dog or sometimes in cat feces and grow in the soil. Skin contact with the parasite is enough to cause infection. Children and people who live in warm, humid climates, such as the American Southeast, are more susceptible. This worm's habits sound like something lifted out of a grade B horror movie: it drills into the skin and then migrates between skin layers, advancing an inch or two a day and leaving red marks and puss-filled blisters as it travels. If the person has been lying on the ground these lesions can be all over the body, and wherever they appear they itch intensely. Fortunately, the larvae do not grow in humans, and after several weeks of torturing their victim, they die and the patient gets better. There are some medicines that help relieve the symptoms (thiabendazole administered orally or topically, ivermectin, or albendazole). In short, this organism is one more reason, if another is needed, not to touch dog feces or let it touch you.

These diseases are not, however, a reason to get rid of the dog. Dogs and cats can and should be treated for both hookworm and roundworm. Deworming of puppies and kittens and their mothers is the best procedure, because they're most likely to harbor the parasites. You should have your puppy treated immediately after birth and then again at two, four, six, and eight weeks of age. Kittens are less often infected, and when they are they are usually infected at an older age. Treatment of kittens can start at about six weeks, then be repeated at eight and ten weeks. There are many safe and effective drugs a veterinarian can use, and you can also administer them yourself at home. You don't need to wait for a diagnosis—the worms are so common in puppies and kittens that all of them should be treated prophylactically.

Bird Trouble

You may remember *Chlamydia trachomatis* from the last chapter—it's the bacterium that causes the most common sexually transmitted disease in the United States, chlamydia. A closely related species, *Chlamydia psittaci*, infects pet birds (and more rarely certain wild birds as well)

and can be transmitted to humans, who can get flulike symptoms from it, and sometimes more serious health problems. Usually the infected person gets a sudden onset of fever, chills, headache, muscle pain, and malaise. In some cases it can also infect other organ systems—the heart, the liver, the eyes, and the brain. It can even be fatal, although it rarely is. It can also cause fetal death in pregnant women. And your pet bird doesn't even have to look sick to be infected with it.

Avian chlamydiosis is one name for the disease this germ causes. Sometimes it's called parrot fever, parrot disease, or psittacosis, after the psittacine birds most often infected—parrots, macaws, cockatiels, and parakeets. But psittacosis can also infect pigeons, doves, budgerigars, and mynah birds. People most at risk are bird owners, pet store employees, and pigeon fanciers. Those whose occupation puts them in close contact with birds are also at risk—poultry slaughterhouse employees, veterinarians, laboratory technicians, farmers, and zoo workers.

Caged birds are the ones that typically get infected, not wild birds. This is probably due to various factors including stress from shipping, overcrowding, chilling, and breeding. The bacteria are shed in the bird's feces and nasal discharges, and they're very durable—they can live on environmental surfaces and remain infectious for several months. You can get it just by breathing it in: the most common way psittacosis is transmitted is by inhaling the organism in dust from dried feces or respiratory secretions. Sometimes people will come down with the disease and not recall ever having had any contact with birds— probably because a few whiffs near a birdcage was enough to make them sick. You can also get it by mouth-to-beak contact and by handling infected birds' plumage.

Spring is birthing season, and the spring of 1996 was particularly busy on one ranch in Montana. Out of a flock of 24 sheep, 19 of them gave birth to 30 lambs over a period of about a month, and a 25-year-old pregnant sheepherder assisted on almost all of the births—on one ewe's premature delivery she spent an hour manually extracting the retained placenta. All the activity calmed down by the end of April, but then, 19 weeks pregnant, the rancher came down with a cough and congestion that persisted for two weeks, followed by four days of fever of 104°F, headache, fatigue, backache, and abdominal pain. She went to the hospital, where she had a chest X ray that looked normal, but she was anemic, had a low platelet count (indicative of infection), and

her liver wasn't working right. They took blood, urine, and sputum cultures, but they couldn't find any germs. By the second day of hospitalization, her infection and fever were so bad that they had to terminate the pregnancy to save her life, but the following day she still had respiratory distress bad enough to require oxygen. A chest X ray at this point showed infection, and antibiotics were started. Within three days she was better and left the hospital. Her discharge diagnosis was "early severe and low platelet count syndrome" with a possible concurrent viral infection.

This diagnosis was not completely satisfying to her obstetrician, who felt it didn't completely reflect all of her symptoms. So he took some placental and fetal tissue samples and sent them to California and Oklahoma for specialized pathologic examination. The lab did a genus-specific fluorescein-tagged monoclonal antibody staining and found masses of *chlamydia*, but they couldn't figure out what species it was. So they sent the stuff to the CDC, where they did a microimmunofluorescence assay, an even more elaborate test, which identified *C. psittaci*. Tests for other kinds of chlamydia were negative. The woman had never been near any birds of any kind, and when they tested the sheep's sera for *C. psittaci*, they found none except in two animals. One of them was the ewe from which the patient had extracted the retained placenta.

It rarely happens, but there have been a handful of reports worldwide of cases of *C. psittaci* infection in sheep, goats, and cattle—they often cause abortions in pregnant animals, and they can sometimes infect humans in close contact with them. Yet another strain infects cats, causing a runny nose and eye infections in the animal, but this form is very rarely transmissible to humans. Humans can't get catch parrot fever from each other, either.

Parrot fever can make a bird ill or kill it, but just as often the animal is asymptomatic. Some birds die soon after they are infected. They get lethargic, can't eat, have ruffled feathers, and produce green droppings and eye and nasal discharges. Other birds can be infected without showing any symptoms at all, and then show symptoms years after exposure. Veterinarians can successfully treat sick birds with antibiotics.

Often, three days to several weeks can pass between the time a human is infected and the time any symptoms appear. If you have symptoms and know you've been exposed to birds, you can have your blood tested for the germ, but treatment even without the blood test is often required. The problem is that the usual tests can't distinguish

C. psittici from other species of chlamydia such as *C. pneumoniae* and *C. trachomatis* (the one that causes venereal disease). There is that elaborate test mentioned above that can make the distinction between species, but very few commercial laboratories are equipped to perform it.

Treatment is relatively simple: tetracycline, a common and inexpensive antibiotic, taken orally for about two weeks. It's important to keep taking the antibiotic for the full two weeks, even after the fever has subsided, because relapses can occur. Seriously ill patients may need intravenous antibiotics. Pregnant women and kids under nine years old shouldn't use tetracycline (it can harm the teeth of kids and unborn babies), so erythromycin is usually the choice for them.

The most common cause of bacterial diarrheal illness is not *salmonella* or *E. coli*, the famous germs, but *Campylobacter*, a not-so-famous one. The CDC estimates that 2.4 million people—almost 1 percent of the population—get the disease every year. It isn't commonly fatal, but it makes a lot of people very sick and a small number of them, mostly very young, very old, or immunocompromised, die from it. It's much more common in the summer than in the winter, infants and young adults get it more than others, and males get it more often than females. There are 16 species of this S-shaped bacterium, one of which, *C. jejuni*, accounts for most human illness. Usually people recover from the illness without treatment, though often only after a week or more of suffering with it. In rare cases, there are long-term consequences. Some people get arthritis after they've recovered from the acute symptoms, and others get an illness called Guillain-Barré syndrome, a disease of the nerves that can require intensive long-term care. Complications such as blood, pancreas, and liver infections and abortion have also been reported, though rarely. (See chapter 3 for more details.)

These bacteria occur in most warm-blooded animals, and the main route of transmission is by eating undercooked food. Even a tiny number of bacteria can cause disease—500 organisms, or the contents of one drop of infected chicken juice, is enough. Cats and dogs can also carry it—they almost never get sick from it, so there's no easy way to tell if a pet is infected. Cats and puppies are more likely to be infected than adult dogs, and they can transmit the disease to humans in their feces.

No one really knows whether the germ is transmitted more often from pets or from contaminated food, but both make a contribution. If a disease occurs in large outbreaks, it's much easier to tell what the

source is. For example, if 30 people get sick and all of them ate the potato salad, you can guess that the potato salad is the problem and then look for a germ in it. But with *campylobacter*, there are rarely large outbreaks. Instead, there are lots of individual, unconnected cases. Figuring out the exact source of the infection under those circumstances is very difficult and often impossible.

Here's some more advice you've probably heard elsewhere: wash your hands thoroughly after changing the cat litter or cleaning up after the dog.

Fishy Business

Well, okay, so I can get plenty of diseases from those pets, but at least I can't catch anything from my tropical fish, right? Not so fast. It would be absurd to tell you that keeping pet fish is a big risk for human illness—it isn't—but that doesn't mean they can't make you sick under certain circumstances. It's not easy to catch anything from a fish tank, but it can be done. In July 1988 in a community hospital in southeastern Missouri, a 14-month-old girl had watery diarrhea and fever of 102°F. She was treated with an antibiotic and was fine after five days. They cultured her stool, and it was negative for campylobacter, salmonella, shigella, yersinia, aeromonas (a still poorly understood bacteria that may or may not cause illness), and rotavirus. She hadn't traveled more than 80 miles from home. She went to a day care center, but no one else in her group was ill. She drank only water from the municipal system, and had waded in two lakes in the area before she became ill. Nothing in any of this would explain her illness. Then the doctors learned that she stayed at a baby-sitter's house once a week, and that the baby-sitter had a freshwater aquarium in which she kept piranhas. The baby-sitter routinely bathed the child before her parents picked her up. You may have already guessed what else she used the bathtub for: cleaning her aquarium. When they tested the water in her aquarium, they found it was contaminated with *Plesiomonas shigelloides*, a bacterium that causes intestinal illness in humans. A test of the baby's stool found the same organism. The Missouri health department was a bit curious about this development, so they took samples from 18 fish tanks in various parts of the state. In four of them they found *P. shigelloides*.

Let's not panic. This is a very rare finding—none of the pet store employees where they found the polluted tanks complained of any illness, and the illness that *P. shigelloides* causes isn't very serious anyway. Still, it wouldn't hurt if you washed your hands after touching the water in your fish tank. And please resist the temptation to bathe in discarded aquarium water.

As we said at the start, don't give up your pets. The benefits of pet ownership far outweigh any health risks in owning them. But a little knowledge—and the exercise of a little caution—can go a long way toward making living with pets a much more appealing proposition.

8

Up Your Nose

THE FLU AND THE COLD

In 1918, the flu season came on a bit early, in April, but a few weeks before that 84,000 additional American troops had set out for Europe to fight the Germans, so there were more important things to worry about than the "grippe." That some people died during the spring was unremarkable—before there were any antibiotics, people often died of bacterial pneumonia, especially in the spring. While records show that men in close quarters—soldiers, for example, and prisoners—got the flu in large numbers, the extent of the general public's affliction went unreported, and therefore largely unnoticed. No one was collecting statistics. In any case, most of the men who got sick were ill for a few days and then got better. There was an odd thing about this flu in addition to its early arrival: it seemed to be affecting most severely people in their 20s and 30s, unlike the usual pattern of serious illness in the very young and the very old. In any case, by August, the Office of the Surgeon General reported that the only serious health problem was that deaths from respiratory disease were up. It was only a slight rise, although it was odd for August, and especially odd that respiratory illness deaths were also up in Europe at exactly the same time. But the flu seemed to have faded. Unfortunately, it would return in the fall, and when it did, it would come back in a way that almost no one in the world could ignore.

In many cases, it began with a dull headache, rapidly followed by chills so severe that nothing could keep the patients warm. The fever led to delirium. Patients began coughing up blood, and their faces

turned purple. Saliva pink with blood bubbled from their mouths as their lungs filled with fluid. Finally they died, essentially drowning, gasping for air. All this could happen in a few days or as little as a few hours. In others, death took longer. For these people, the fever lasted four or five days, at which point bacteria invaded their damaged lungs, bringing death from bacterial pneumonia. It was like no flu before or since. By the time it was over, an estimated 20 to 25 million people worldwide were dead from the virus. And the most terrifying part was that this flu, unlike any other, killed people in their prime as well as infants and the elderly. Adults between 20 and 40 were among the most susceptible age groups: 40 percent of the American navy and 36 percent of the army came down with the disease. In all, 28 percent of Americans were afflicted, and 675,000 of them were dead by March 1919. This flu killed with a relentlessness and rapidity unequaled by any plague in history. (Mortality figures are complicated by many factors, and deaths from a given disease must necessarily be an estimate. There were no serology tests for the flu virus in 1918.)

Doctors in 1918 had the foresight to preserve tissue samples from flu victims, and these samples have now been analyzed using modern laboratory techniques. In one experiment, the virus was reproduced in tissue culture and injected into mice. Although it did not have the same devastating effect on the mice that it had on humans, its viral RNA has been sequenced, showing consistency with an H1N1 influenza A virus that belongs to a subgroup that infects humans and pigs. The pandemic of 1918, in other words, may have been a kind of "swine flu."

Flu

Although you might not think of the flu as a killer disease, an average of 20,000 Americans die of it every year—that's about four times as many people as die from the diarrheal illnesses caused by eating contaminated food. Yet many people who are sure they are going to suffer an agonizing death from eating potato salad contaminated with salmonella view the flu as a sort of mild and unavoidable annual unpleasantness.

Most of the people who die from the flu are older than 65 or younger than 1 year. In the ages in between, deaths usually occur from pneumonia (either primary viral pneumonia or bacterial pneumonia as one of the complications of influenza infection) and from exacerbations

of cardiopulmonary conditions or other chronic illnesses. There are an estimated 114,000 hospitalizations caused by flu infection yearly, and 43 percent of them are people younger than 65.

The flu is caused by two different kinds of virus called influenza A and influenza B.[*] There are numerous subtypes of influenza A, but influenza B viruses are not categorized into subtypes. These viruses have RNA rather than DNA as their genetic material. When an RNA virus reproduces, it doesn't make an exact copy of itself like the DNA cells in our bodies do. Instead, "errors" are constantly being introduced. Poliovirus is another example of an RNA virus—but, fortunately, it's an exception to the general rule. It has only three different types, and that's one reason why the vaccine against it has been so successful. Once the polio vaccine gives us immunity, we are immune to polio. Not so with the flu.

There are two kinds of changes that occur in influenza A viruses. The first is a series of mutations that happen over time, causing a gradual change in the virus that makes it able to avoid your immune defenses so that you're susceptible throughout your life. This gradual change is called antigenic drift, and it gives the virus great staying power—a vaccine effective against one generation is no longer effective against the next. Because of this, new vaccines are needed every year to combat "this year's strain" of flu virus. The other kind of change is more abrupt and involves changes in the hemagglutinin and neuraminidase surface proteins of the virus, the proteins that allow the virus to infect a cell, and to which the body's immune system reacts. This change, called antigenic shift, suddenly produces a new subtype. While type A viruses can undergo both kinds of changes, type B viruses only demonstrate the more gradual antigenic drift.

These rapid changes in the genetic material of flu viruses allow them to infect animals as well, and the virus may be of animal origin in the first place. Some of us remember the swine flu scare of the late 1970s, which turned out to be less of a threat than first thought. This flu virus developed from pigs that were infected simultaneously with a pig and a human form of influenza A. The two viruses exchanged genetic information in the pigs, creating an entirely new strain that was infectious to humans. In 1997, an influenza outbreak began killing chickens

[*]There is a third type called influenza C, but it usually causes mild symptoms or none at all. It is not a cause of epidemics, and has no severe public health impact.

in markets in Hong Kong. The virus was a type of influenza A called H5N1. An "H5N1" influenza virus means that its surface proteins are hemagglutinin type 5 and neuraminidase type 1. The government of Hong Kong immediately closed all retail and wholesale markets selling chickens and destroyed more than 1 million infected birds. The vigilance of the Hong Kong government was fully warranted: 18 people in Hong Kong were in fact infected, and six of them died from complications of the disease. Destroying the chickens seemed to banish the virus—but not permanently. The bird flu returned twice in 2001, with the government ordering the destruction of chickens, which had come from China, both times. Researchers now believe that a change in one particular gene of the virus that is used in viral replication is what allowed the disease to move from birds to humans. Apparently pigs are not necessary as an intermediary in the development of pandemic strains of the flu.

This Year's Flu Vaccine

How do we find out about new strains in time to develop new vaccines against them? The World Health Organization, in cooperation with dozens of other national and international health groups, maintains the Global Outbreak Alert and Response program to provide continuous surveillance of infectious disease outbreaks around the world. This program covers all kinds of infectious disease outbreaks, and there are various specialized surveillance programs, including FluNet, a geographical system that monitors influenza activity. There are 110 National Influenza Centers in 82 countries, and four Collaborating Centers for Reference and Research, including one at the CDC in Atlanta. Each of the 110 national centers collects information and sends virus isolates to one of the four research centers for strain identification. (If ever an argument for international cooperation were needed, the activity of infectious microorganisms provides it: preparing for and responding to epidemic disease can be done effectively only with an internationally organized effort. Viruses are immune to immigration laws.) On the basis of this information, WHO arrives at its conclusions about what should be included in this year's flu vaccine and then makes its recommendations; individual countries and their vaccine manufacturers are responsible for carrying out the vaccination programs.

In June 2002, for example, the FDA recommended the contents of the 2002–2003 vaccine for the Northern Hemisphere. That winter's vaccine contained three components: A/New Caledonia/20/99-like (H1N1), A/Moscow/10/99-like (H3N2), and B/Hong Kong/330/01-like viruses, becuse these three types were the most likely to strike that winter. (The name of the city tells you where the virus was first isolated.) The flu vaccine is protective for up to 80 percent of the people who take it and encounter one of these strains. It is, in addition, partially effective for other strains, in that it helps make the disease less severe and less deadly.

The flu is quite contagious and is transmitted from one person to another primarily by coughing and sneezing. Between one and four days pass from the time you're infected until you begin to show symptoms—two days is about the average incubation period for the flu virus. You can infect others from about one day before symptoms begin until about five days after they start. With children, the period of infectivity can be longer. You're probably quite familiar with the symptoms. There is an abrupt onset of fever, muscle ache, headache, and malaise. You get a nonproductive (dry) cough, and sometimes a sore throat and a runny nose. The symptoms are difficult to distinguish from the symptoms of other respiratory diseases, and the only sure way of knowing you have the flu is by getting an antigen test or a lab culture done. Most otherwise healthy people don't bother with this because in a few days they feel better anyway.

There are some antiviral medicines, which we'll get to in a moment, but their effectiveness is limited. The only really effective treatment is prophylactic—that is, the flu vaccine. If someone tells you—and someone probably already has told you—that he got the flu from the flu vaccine, he is mistaken. You cannot get the flu from the flu vaccine. There are three different kinds of vaccine. One is prepared from killed (inactivated) virus that has been grown in hen's eggs and is highly purified; the two others contain no virus at all, only the part of the virus that induces an immune response (these are called the purified-surface-antigen preparation and the subvirion preparation, and sometimes both are referred to as "split virus" preparations). None of these can cause the flu. This does not mean, however, that you cannot get the flu after you have had the vaccine. The vaccine is, as we said above, about 80 percent effective—and that's against the viruses it is designed to combat. It's largely ineffective against other strains. So you might get the flu

after you have the vaccine and then mistakenly attribute your flu to the vaccine. You did not, however, get the flu *because* you had the shot. Very rarely, the inactivated vaccine can cause a fever—but it's not the flu. For this reason the subvirion and purified-surface-antigen preparations, which are less likely to provoke this reaction, are recommended for children. (In the 2001–2002 season, the only vaccines distributed in the United States were of the split virus type.)

Should you get the flu vaccine this year? The CDC recommends that any person over the age of six months who is at increased risk for complications from the flu should get the shot. This means you should get the flu vaccine if you:

- Are over age 50.
- Are a resident of a nursing home or any other chronic care facility.
- Have a chronic disorder of the pulmonary or cardiovascular system (including asthma).
- Have required regular medical follow-up or hospitalization during the previous year for chronic metabolic diseases (including diabetes), renal dysfunction, hemoglobinopathies, or immunosuppression.
- Are immunocompromised, including immunosuppression by HIV infection or medications.
- Are a child or teenager receiving long-term aspirin therapy, and therefore might be at risk of developing Reye's syndrome as a complication of the flu.
- Are a woman who will be in the second or third trimester of pregnancy during the flu season.
- Work in a hospital, an outpatient care setting, or as part of an emergency response team.
- Work in a nursing home, an assisted living facility, or other residence for people at high risk.
- Provide home care to people at high risk.
- Have any household members (including children) who are at high risk.

The CDC has this to say about vaccination of pregnant women: "Because currently available influenza vaccine is an inactivated vaccine, many experts consider influenza vaccination safe during any stage of pregnancy. A study of influenza vaccination of more than 2,000 preg-

nant women demonstrated no adverse fetal effects associated with influenza vaccine. However, more data are needed. Some experts prefer to administer influenza vaccine during the second trimester to avoid a coincidental association with spontaneous abortion, which is common in the first trimester, and because exposures to vaccines have traditionally been avoided during this time." (Centers for Disease Control and Prevention. *Health Information for International Travel 1999–2000,* DHHS, Atlanta, GA.)

There are certain people who should not get the vaccine. The inactivated vaccine should not be given to anyone with a known hypersensitivity to eggs or any other component of the vaccine without first consulting a physician who may recommend you get the shot after desensitization for the allergy. (This does not apply to the split virus vaccines, which have no egg component.) People who have an acute illness with a fever should not get the shot until the fever has abated. But minor illnesses with or without fever, especially among children with mild upper respiratory infections, are not a contraindication for the vaccination.

When should you get the vaccine? The best time is in October or November, but vaccination after December is still a good idea, since the peak of the flu season can be as late as March.

Although flu season is October to March in the northern hemisphere, it isn't everywhere. In the tropics, it occurs all year round, and in the southern hemisphere the flu season is April through September. If you're traveling in a large group of people—say on a cruise ship—you can get the flu any time of year, no matter where you are, because there are likely to be fellow travelers who come from parts of the world where the flu virus is currently circulating. The flu can, at best, mess up your vacation. At worst, it can cause serious complications. It's best to get the current vaccine if you're planning a trip. If you weren't vaccinated in the fall preceding your trip, you should consider getting the shot at least two weeks before you travel—but remember that the availability of the vaccine in the summer is likely to be limited. The benefits of revaccination before travel of people who were vaccinated the previous fall or winter are not known, but current practice is to give high-risk nursing home residents two shots, one in September and one the following January.

There are nasally administered live, attenuated influenza virus vaccines that have been in development in the United States since the

1960s and are presently being used in Russia. These are active viruses that cause minimal symptoms but provoke a protective immune response. Recent studies have found them effective in preventing flu and reducing otitis media in children. These vaccines are not currently available in the United States but may be soon, perhaps even by the time you read this.

There are also some antiviral medicines that are a useful addition to the vaccine in preventing and treating influenza, although they are not a substitute for the vaccine. Amantadine, rimantadine, zanamivir, and oseltamivir have been approved for treatment and prophylaxis of influenza. The drugs have varying side effects and costs. If you take them within two days of the onset of illness, they can reduce the duration of illness by about one day. All four drugs are effective against influenza A. Zanamivir and oseltamivir are somewhat effective against B as well.

These antiviral medications can be useful in special circumstances, such as protecting high-risk people after the flu season has begun, protecting high-risk people who for one reason or another can't take the vaccine, or protecting the immunocompromised who don't get a sufficient immune response from the vaccination alone. They also have some prophylactic effect in that they can confer immunity by allowing subclinical illness while the body develops protective antibodies against the virus, and they do not interfere with the immunologic response to the vaccine. All of these drugs have side effects, often much more unpleasant than any side effects of the vaccination. Zanamivir is difficult or impossible for some people to take—it is given as an inhaled powder, and those with asthma or other lung ailments can't use it.

Antibiotics are of no use in treating influenza and should never be taken for that purpose. Unfortunately, they often are demanded by patients and prescribed by doctors or other health care professionals. In one survey, fully 60 percent of Medicaid patients were prescribed antibiotics for viral illnesses. This is bad practice—if these patients had nothing more than the flu. The problem is that the symptoms of flu are sometimes very difficult to distinguish from those of bacterial illnesses for which antibiotics are indicated, and sometimes a bacterial illness can accompany the flu, as was frequently the case, with fatal results, in the 1918 pandemic. There are tests that can determine with certainty whether the flu is the cause of the illness, including some that can pro-

duce results in as little as 30 minutes. The tests vary in their ability to distinguish different virus types, and the rapid tests are less reliable than the viral culture and blood tests, which take much longer. Information about the exact strain may not be necessary in treating individual cases, but such information is essential for tracking epidemics and developing new vaccines.

Despite considerable knowledge of the genetic structure of the 1918 virus, no one is quite sure why the flu that year was so virulent, and there is still considerable scientific disagreement about where it came from. It is likely that a new surface protein appeared on this virus, one that humans had never encountered before and to which they therefore had no immunity. But there must have been other genetic features of this virus that also contributed, and these remain a mystery, even though complete viral gene sequences of the bug have been developed from the preserved tissue. So the question arises: Could it happen again? In a sense, it already has happened again, though in not quite so devastating a form. The "Asian flu" of the winter of 1957–1958 killed 70,000 Americans. The 1968–1969 "Hong Kong flu" killed 34,000. The new subtype that appeared that year, called A(H3N2), was milder, perhaps because only the hemagglutinin protein changed—the neuraminidase stayed the same, and therefore people may have had some residual resistance to it. Still, A(H3N2) has caused 400,000 deaths in the United States since its emergence in 1968, 90 percent of them among people older than 65. In fact, the subtype that caused the 1918 pandemic, A(H1N1) continued to circulate for years, having undergone antigenic drift, and caused a large number of deaths in 1920 (reduced, probably, by greater immunity among the population as well as by the attenuated form of the virus), and then periodic outbreaks until it disappeared in 1957. None of the outbreaks, though, was anywhere near the severity of that of 1918. In 1977 it reappeared, but again without devastating consequences.

In any case, most virologists think that the answer is yes, it could happen again. The reservoir of influenza is believed be wild waterfowl, and the pandemics of 1957 and 1968 were both produced by changes in the surface proteins of an avian strain of the virus. Such an alteration in the future could produce a pandemic as bad or worse than the 1918 disaster. Of course, circumstances now are a bit different. We have a flu reporting system; we know how to make vaccines against the flu; we have antibiotics to treat the bacterial infections that killed so many in the pandemic

of 1918; we even have several antiviral medicines that are helpful. But we also know that microbes are persistent and very adaptable, and that there is always the possibility that one of them will come up with an answer to the most elaborate defenses we have constructed against it.

Colds

If we want to search for something good to come out of the 1918 disaster, perhaps we can find it in the spur to research in virology that it provoked. In the 10 years after 1918, thousands of books and articles were published about the flu. One large problem posed by the flu virus is that it is so small. Scientists in these years had no way of seeing it. They could see many microorganisms, of course, but all of them were the much larger bacteria that cause diseases associated with the flu, but not cause the flu itself. In addition, viruses, unlike bacteria, are very difficult to culture in a petri dish—they normally only live parasitically inside host cells.* On top of this, the flu virus is fussy about its host. The kinds that infect humans don't often infect animals in the same way, and humans, for many and quite obvious reasons, make very poor experimental subjects. Despite all this, by the early 1930s several different viruses that cause the flu had been filtered out from the pool of microorganisms found in flu victims, and the viral nature of influenza was clearly established. In addition, the distinction between influenza and the relatively harmless cold was now becoming clear.

You are even more likely to get a cold than to get the flu—you and 62 million other Americans this year will get at least one. Most of you—52.2 million—will be under 17 years old. There is no vaccine to pre-

*There is a story here worth telling. The title of Alexander Fleming's 1929 paper announcing his discovery of penicillin was "On the Antibacterial Actions of Cultures of a Penicillium, with Special Reference to Their Use in the Isolation of B. Influenzae." Studying the organism that caused death from influenza was a major preoccupation of microbiologists at this time, and Fleming viewed penicillin as useful in that quest—that is, it could, in a petri dish, inhibit the growth of certain bacteria, making the organisms that it didn't inhibit easier to find and study. In other words, his chief hope for penicillin was that it could be used by researchers to eliminate the presence of various bacteria on a culture plate, leaving the organism thought to cause influenza unharmed. He referred to its possible use as an antibiotic in treating humans only briefly at the end of the paper.

vent it; there is no medicine to cure it. This is not to say that people haven't tried. In fact, there is a vast literature of cold cures, and some are honestly convinced that one or another of them works. It is not too cynical, however, to add that there are some who are willing to advertise and sell cures that they know full well are ineffectual.

Since colds are very common, and largely untreatable except symptomatically, there is an irresistible desire to cite just about anything as a cure, or at least a good treatment. Onions, garlic, dill, wild indigo, cinnamon, teas and soups of all kinds, echinacea, hibiscus flowers, Peruvian balsam, marjoram, anise, spruce, peppermint, cardamon, purple cornflowers*—if you can name it, it has been recommended by someone somewhere as the best thing to take if you have a cold. Some of these remedies come with no promises or only very vague ones; others, like the ingestion of vast quantities of vitamin C, come decorated with scientific-sounding explications and endorsements from famous people. Undoubtedly, some of them will make some people feel better, but this is not the same thing as saying that they cure colds. Perhaps the best we could say of all these medicines is that they will work to cure colds about as well as anything else, which is to say they probably won't work at all. You can take them, and most of them (though perhaps not all) are harmless, but you'll still get a cold, and you'll still have to wait until it goes away. In the vast majority of cases, if you're really careful; take vitamin C; use aspirin, saline nose spray, and other cold remedies; drink lots of liquids; and get plenty of rest, your cold will last about seven days. If you don't, it will last about a week. Rarely, a cold can lead to complications, particularly bacterial infections that can cause sinusitis, bronchitis, and bacterial pneumonia, diseases that may require treatment with antibiotics. Colds themselves do not respond to antibiotics, which should never be used for that purpose.

Colds are caused by more than 200 different viruses, and about a third of them are rhinoviruses—nose viruses, in plain English, but of course cold (and flu) viruses specifically infect the lungs. The nose provides the most obvious symptoms—sneezing and dripping. Fever, headaches, and muscle aches are also common. Rhinoviruses, members

*This random sample is taken from the *PDR for Herbal Medicines,* 1st ed. (Medical Economics Company, Montvale, N.J., n.d.). The book prudently lists in an appendix the names of poison control centers in every state and the District of Columbia.

of the genus *Picornavirus*, seldom cause serious illness. Other cold viruses such as parainfluenza and respiratory syncytial virus cause mild disease in adults, but more severe illness in children. Another genus of viruses, the coronaviruses, probably also cause a large number of mild colds in adults, and other viruses can be involved as well. Between 10 and 15 percent of colds are caused by viruses that can also cause more serious illness—adenoviruses, Coxsackie viruses, influenza viruses, and others. But that still leaves about half of all colds for which no one knows the exact (presumably viral) cause.

Colds are usually a winter disease in the northern hemisphere. Although there is considerable controversy about exactly why certain diseases have seasonal patterns, there is no evidence that, say, going outside with wet hair on a cold winter day makes you any more susceptible to colds than if you'd dried your hair thoroughly before venturing out. Many diseases do have seasonal patterns, including flu, rotavirus, measles, and polio, among others, and the patterns are remarkably consistent from year to year. Latitude correlates very closely with the patterns—the magnitude of seasonal peaks increases proportionately with distance from the Equator in either direction.

Since these seasonal cycles are so obvious and universal, and since no one has ever come up with a completely satisfactory reason why they happen, there are plenty of explanations offered by scientists and laypeople alike. A favorite layman's explanation is that being cold gives you a cold, or, slightly more technically, being cold makes you more susceptible to the virus's effect and reduces your resistance. The problem is that no experiment designed to test this hypothesis, and there have been many of them, has ever proven any such causal connection. Some people feel that viruses migrate, that they travel back and forth to cooler (or in the case of certain viruses like polio, warmer) climates as the weather changes, or that they prefer rainy or dry weather and move with the seasons on that basis. Correlations with the weather or temperature and the magnitude of a given pandemic can be made, but the problem is that the same virus that correlates with wet or warm weather in one place has an equally convincing correlation with dry or cold weather in another. Respiratory syncytial virus epidemics, for example, correlate with the cold months of winter in the United States, but in Hong Kong the same viral epidemics are closely correlated with the hotter months. Some say it's the seasonal behavior of the human host—crowded into small rooms in the winter, swimming in public swimming pools in the summer, for example—that causes the seasonal

changes in infection rates. But this doesn't explain why measles spreads in the winter in crowded rooms in England while failing to spread in the summer in crowded rooms at international conferences. Even the light/dark cycle and the corresponding pattern in the secretion of melatonin has been invoked as a possible explanation. There are plenty of guesses, but no one really knows for sure why there are winter and summer diseases.

Why you catch cold in the winter is still a mystery, but how you catch one is not. You catch it by having a virus get into your nose. You take a breath too close to someone who has just sneezed. More commonly, you shake hands with someone with a cold who has just scratched his nose and then you scratch your own nose, or you touch a door handle just touched by a cold sufferer and then put your hand near your face. You don't even have to touch your nose—rub your eye and tears from the eye drain through the nose, giving you a cold. One experiment showed the hand-to-hand transmission of colds to be extremely effective. This study found that 11 of 15 hand-contact exposures of infected people with uninfected volunteers transmitted the illness. Then they tested how efficient this kind of transmission might be and found that the virus on infected volunteers' hands was transmitted to recipients' fingers during 20 of 28 ten-second hand contacts. That's not a very long contact, and the distance from hand to nose, as we know, is quite short. The "hand-contact-followed-by-self-inoculation" route is very likely an important way in which the virus gets around.

The symptoms of a cold—the runny nose, the coughing, and the sneezing—are probably, like the symptoms of many other diseases, a reaction by the body's immune system to attack by a foreign invader. White blood cells called to the site of infection by chemical signals from the virus-infected cells give off a number of different chemicals to try to fight the infection. It is these chemicals that probably cause the swelling and inflammation of the nasal membranes and the leakage of fluid from capillaries and lymph vessels, with the resultant increased production of mucus. These chemicals are of great interest to researchers, who are trying to figure out whether drugs to block them might help alleviate cold symptoms.

Cold Comfort

Since we know how you get colds, we also know how you can prevent them, or at least minimize their number. Stay away from people with

colds; wash your hands after touching someone who has a cold or something they've touched; keep your hands away from your face. If you have a cold, cover your mouth when you cough or sneeze, preferably with a tissue and not your hand. Stay away from people who have asthma or other kinds of lung ailments during the first three days of your cold, because that's when you're most contagious. Don't smoke or breathe secondhand smoke—you don't catch cold that way, but smoke makes colds worse. It is, obviously, not always practical or possible to follow all of this advice.

Once you have a cold, there's nothing you can do to make it go away faster. There are those more or less effective antiviral medicines described above, but none of them works for cold viruses. Still, you can take various medicines that will provide some relief from the symptoms, and drinking plenty of liquids will keep you hydrated and therefore keep the mucus flowing out of you. Acetaminophen— Tylenol—can help you feel less achy and more comfortable, and is less likely to give you a stomachache than aspirin, which can also be used, but not for kids under 18. (Researchers have found that aspirin increases the amount of rhinovirus shed in nasal secretions, which may make you more of a hazard to others than if you don't take aspirin.) Decongestants and antihistamines can help unblock a stuffed nose and relieve a cough, although the newer nonsedating antihistamines aren't as effective as the ones that make you sleepy. None of these will shorten the time it takes to get better.

Some people think that inhaling steam is a good treatment for colds, assuming that increasing the temperature of the nasal membranes will inhibit the replication of virus. Steam may temporarily relieve the symptoms of congestion, but studies have demonstrated that it has no effect on other cold symptoms or on the amount of viral shedding in people afflicted with rhinoviruses. Researchers have widely studied interferon-alpha as a cold cure and given in daily doses by nasal spray, it does appear to have an effect in preventing illness. But interferon causes unacceptable side effects, including nosebleeds, and does nothing for a cold already in progress.

How about a vaccine for the cold? Not very likely. With more than 200 viruses that can cause colds, a single vaccine can't do much. Add to that the genetic changes that all viruses undergo, and you have what appears to be an insurmountable barrier to the creation of an effective vaccine. There is a continuing effort, however, to create an

antiviral medicine that will work against cold viruses. Most recently, one called pleconaril has been developed. It appears to inhibit the reproduction of many different rhinoviruses. The company that makes it, a small pharmaceutical firm called ViroPharma, has been testing the drug in clinical trials for some time and is almost ready to seek approval from the Food and Drug Administration to manufacture the medicine.

Kids under five have the most colds of any age group, probably because they have developed few immunities to the cold viruses, and more boys under five get colds than girls. By the time they're older than five, girls have colds slightly more often than boys. This slight sex difference persists throughout life, with women having slightly more colds than men. Teenagers and adults have colds about two to four times a year; younger kids have them six to eight times.

The American Lung Association would like to disabuse you of some received wisdom about colds, and we see no reason not to help them:

- Vitamin C has not been proven to prevent or cure colds.
- Zinc lozenges, gels, and nasal sprays have not been proven to cure colds.
- Exposure to extreme cold does not lead to colds (or the flu).
- Going to work or school with a cold will not prolong the symptoms.
- "Starve a cold and feed a fever" is not good advice. You should increase your intake of fluids when you have a cold, with a fever or not, and eat enough to satisfy your appetite.
- There is no evidence that herbal treatments like echinacea work to cure colds.
- Alcohol, in the form of a hot toddy or otherwise, is not good for a cold and should be avoided.
- Chicken soup possesses no curative properties. It does, however, taste good, and liquids are good for you when you have a cold.

Do you just have a cold, or is it the flu? Although you can't know for sure without a viral culture or serology test, the symptoms differ, and you can usually make an educated guess. The National Institute of Allergy and Infectious Diseases has put it in a table.

Cold or the Flu?

Symptoms	Cold	Flu
Fever	Rare	Characteristic, high (102°–104°); lasts 3–4 days
Headache	Rare	Prominent
General aches, pains	Slight	Usually severe
Fatigue, weakness	Mild	Can last up to 2–3 weeks
Extreme exhaustion	Never	Early and prominent
Stuffy nose	Common	Sometimes
Sore throat	Common	Sometimes
Chest discomfort, cough	Mild to moderate; hacking cough	Dry cough common, and can become severe

Although research continues into promising treatments for colds and vaccine technology is constantly improving, we have not in any sense "conquered" the viruses that cause colds and flu. They are now, and will continue to be for the foreseeable future, our constant unwelcome companions.

9

Bottled or Tap

WATER, AND WHAT'S IN IT

The Contaminant Candidate List sounds like it might be a registry of anti-environmental office seekers published by the Sierra Club. But it's actually a list of the contaminants in public drinking water that may require regulation, and the Environmental Protection Agency is required to issue such a list every five years. The last one came out in 1998, and in addition to dozens of different chemical contaminants, it lists 10 scary microbial organisms that might be found in municipal water systems around the country and might have to be added to the list of contaminants already regulated:

- Acanthamoeba (guidance expected for contact lens wearers)
- Adenoviruses
- Aeromonas hydrophila
- Caliciviruses
- Coxsackie viruses
- Cyanobacteria (blue-green algae), other freshwater algae, and their toxins
- Echoviruses
- *Helicobacter pylori*
- Microsporidia (enterocytozoon & septata)
- Mycobacterium avium intracellulare (MAC)

There are some real gems on this list. *Helicobacter pylori,* for example, causes peptic ulcers. Coxsackie viruses cause a disease in children

that gives them a fever, sores in the mouth, and a rash with blisters. Adenoviruses cause respiratory illness, whose symptoms range from the common cold syndrome to pneumonia, croup, bronchitis, and conjunctivitis. Acanthamoeba can cause a deadly infection called granulomatous amebic encephalitis (GAE). You get headaches, a stiff neck, nausea and vomiting, loss of balance and bodily control, seizures, and hallucinations. Central nervous system infections with acanthamoeba are almost invariably fatal. And this is just the stuff that *isn't* yet regulated. What's tracked, reported on, and regulated, you can be pretty sure, is even worse.

Don't Drink the Water

On September 3, 1999, 10 children in counties near Albany, New York, were hospitalized with bloody diarrhea, and everyone wanted to know why. Health officials quickly discovered two facts: they were all infected with *E. coli* O157:H7, the most deadly form of the *E. coli* bacterium, and they had all attended the Washington County Fair, held two weeks earlier. By September 15, more than 900 fair attendees had reported diarrhea, and examination of stool cultures revealed that 116 of them were infected with *E. coli* O157:H7. Sixty-five people were hospitalized and 11 children had developed hemolytic uremic syndrome (HUS), a disorder that damages blood cells by causing waste products to accumulate in the blood instead of being excreted in the urine. Two people, a three-year-old girl and a 79-year-old man, both of whom had HUS, were dead. Finally, an environmental study of the fairgrounds uncovered the truth: most of the water at the fair was chlorinated and perfectly safe for drinking, but one area where several food vendors used water to make drinks and ice was supplied by a well. The well water was highly polluted with *E. coli*, and careful testing showed that the strain was O157:H7.

Stories like this may make you want to buy bottled water and a water filtration system—and in fact those are two growth industries in the United States. Americans spend nearly $6 billion a year on bottled water—and it's no wonder, considering the way restaurants, convenience stores, and soda vending machines are pushing the stuff. But carrying a little plastic bottle of water, preferably one with a foreign name written on it, is more a fashion statement than a physiological neces-

sity. The fact is that most municipal water systems in this country, and most wells, are perfectly safe, just as safe as a bottle of Perrier. But every once in a while things seem to fall apart, as they did in Milwaukee in 1993, when the failure of the purification system there caused an outbreak of cryptosporidiosis that left 400,000 people sick and 100 dead—but more about cryptosporidiosis in a moment.

When you look at the list of pathogens that have turned up in water supplies over the past few years—giardia, salmonella, campylobacter, yersinia, *Entamoeba histolytica, E. coli,* shigella, cryptosporidium—it's enough to make you reach for bottled water even if you just want to brush your teeth. And we haven't even gotten around to the presence of lead, arsenic, chloroform, pesticides, and radioactive waste in your drinking water, things a little off our main subject here, but which, as you might imagine, come with their very own horror stories. But let's not panic about pathogens just yet. The tale of the turbid water in Washington, D.C., is instructive.

On December 6, 1993, the District of Columbia began to have difficulties with their filtration system, and levels of turbidity (that is, suspended particles in the water) rose above levels allowed by Environmental Protection Agency standards. Turbidity can be caused by many different pollutants, including some things that are completely harmless. But it's aesthetically undesirable, the solids may contain something harmful, and it can decrease the effectiveness of water treatment techniques by protecting pathogens from chemical or thermal damage. D.C. water is chlorinated, and throughout the period of the failing filtration, chlorination remained normal. Chlorine kills many different waterborne organisms, but there is at least one that it doesn't kill: cryptosporidium, a common cause of intestinal disease.

Cryptosporidium is a protozoan parasite discovered in 1976, and it exists all over the world. Along with *Giardia lamblia,* another protozoan, it is the most common cause of waterborne disease in the United States. There are many different species that infect animals, but only one, called *Cryptosporidium parvum,* infects humans. The organism has a complex life cycle, including sexual and asexual reproductive stages, and exists in large quantities in human stool. You get it by ingestion—that is, by getting stool into your mouth. Crypto, as it is called for short, can survive for several days in water, and if you swallow enough of that water, you get a stomachache, bad diarrhea, and sometimes a fever. The unpleasant symptoms can last for quite a while—up to two weeks—and

they usually begin between 2 and 10 days after you're exposed to the parasite. You can die from this infection, as some did in the Milwaukee case, but almost all deaths are among the very old, the very young, and the immunocompromised. In fact, many people may have cryptosporidium in their stool without experiencing any symptoms at all.

In any case, the authorities in Washington were concerned that crypto might be one of the things suspended in the water. The water was eight times more turbid than the water in Milwaukee that made 400,000 people ill—that is, eight times as many particles were suspended in it. But the tests to determine the actual presence of crypto in water are both expensive and highly inaccurate, so most water companies don't use them. DC was luckier than Milwaukee. Apparently, whatever was causing the turbidity in the water in Washington wasn't causing cryptosporidiosis, because reports of diarrheal illness were no higher after the filtration problems started than they were before. The water didn't look very good, and it certainly frightened lots of people, including those in charge of the water system, but in the end, unlike the Milwaukee case, it probably didn't do much harm.

Cryptosporidium parasites are large enough that they can be filtered out of drinking water, and most municipal water systems, like those in Milwaukee and Washington, D.C., when they're working right, do just that. But if you have reason not to trust the municipal system, you can buy your own filter and attach it to your tap or have it installed on the pipes leading to the tap. There are lots of home water filters sold, but not all of them will work to catch cryptosporidium before it winds up in your drinking glass. To find out if a particular filter removes crypto, you can contact NSF International, a nonprofit organization that does voluntary testing of many products (3475 Plymouth Road, P.O. Box 130140, Ann Arbor, MI 48113-0140, 800-673-8010). Any filter that NSF has tested and that works against crypto will have the words "Standard 53" and "cyst reduction" or "cyst removal" on the label. NSF testing is paid for by the product manufacturer, and since it is expensive, not all manufacturers want to acquire this seal of approval. This doesn't in itself mean that the filter will not be effective against crypto, however. There are four phrases you can look for that are sure indications that the filter will work:

- Reverse osmosis (with or without NSF testing)
- Absolute pore size of 1 micron or smaller (with or without NSF testing)

- Tested and certified by NSF Standard 53 for cyst removal
- Tested and certified by NSF Standard 53 for cyst reduction

There are lots of other descriptions that can appear on water filter labels, but none except those above assures you that crypto will be filtered out. Filters that have wording like the following on them may or may not work:

- Nominal pore size of 1 micron or smaller
- One micron filter
- Effective against *Giardia*
- Effective against parasites
- Carbon filter or activated carbon
- Water purifier
- EPA approved or EPA registered (this is particularly deceptive, since the Environmental Protection Agency doesn't test, approve, or register water filters)
- Removes chlorine
- Ultraviolet light
- Pentiodide resins
- Water softener

Changing filter cartridges at the recommended intervals is essential, and anyone who changes the cartridge should wear gloves and wash his hands thoroughly after handling the old filter.

What about other drinks? Can they contain crypto? Maybe. In general, canned or bottled soda, seltzer, and fruit drinks are free of cryptosporidium thanks to the manufacturing and canning process, but fountain drinks, iced tea and coffee, and concentrated juice that you mix with tap water may not be.

You may decide that bottled water is the way to go, but even here, you have to be careful. The water may be in a neat little bottle, but that doesn't mean that it has been effectively treated to eliminate crypto. Again, there are a few phrases you can look for that will assure that an effective method has been used:

- Reverse osmosis treated
- Distilled

- Filtered through an absolute 1 micron or smaller filter
- One micron absolute

Other phrases don't mean much when it comes to cryptosporidium. If you see any of the following on the label of a bottle of water, you can't be sure that crypto has been eliminated:

- Filtered
- Carbon-filtered, particle-filtered, multimedia-filtered
- Ozonated or Ozone-treated
- Ultraviolet-light treated
- Activated carbon treated
- Carbon dioxide treated
- Ion exchange treated
- Deionized
- Purified
- Chlorinated

Deciphering Water Bottle Labels

If the bottle label reads "well water," "artesian well water," "spring water," or "mineral water," that certainly sounds healthful, and may even be healthful, but such sources can be polluted with cryptosporidium, too. There is a large variety of water brands, and an almost equally large variety of sources from which the water comes. A bit of label reading is useful.

Drinking water regulation is the responsibility of the Environmental Protection Agency (as well as state government agencies). But when drinking water is put in bottles and sold, it's a food and therefore comes under the jurisdiction of the Food and Drug Administration. Bottled water has to meet all the EPA's rules about purity of tap water, and it has to be labeled legally. In 1993, the FDA promulgated regulations about the labeling of bottled water. If you see "spring water" on the label, that means that it actually did come from a spring—"a geologically and physically protected underground water source"—and the water has to be collected from the spring or from a bore hole next to the point where it emerges from the ground. Mountain Valley in Arkansas, Belmont Springs in Massachusetts, Saratoga Springs in New

York, and Poland Spring in Maine are well-known sources of such water. If it says "mineral water," it has to have at least 250 parts per million in total dissolved solids, and it, too, has to come from a "geologically and physically protected underground water source." Water bottled from municipal water sources must be labeled as such, but not if it has been processed and treated so that it can be labeled "distilled" or "purified."

"Naturally sparkling" water has to really be naturally sparkling as it comes out of the ground—not made bubbly with artificial carbonation. This has created a minor problem for Perrier, probably the most famous bottled water in the world. At one time, water from the spring at Les Bouillens bubbled up through the ground into a pool, naturally slightly carbonated. But the pool was filled in early in the twentieth century, and now Perrier drills down into the underground reservoir for its water. Unfortunately, the reservoir isn't carbonated. The CO_2 that produces the slight carbonation of Perrier comes from another stratum below, whose gas used to mix naturally with the water but does not do so today. So they drill into that stratum, too, and mix the gas and water in a bottling plant after they've brought both to the surface. This is done in just the right proportions—3.5 grams of gas per liter of water—to produce those "small bubbles" that Perrier is famous for. So the carbonation is "natural," if you want, and it's natural enough for the FDA, which allows Perrier to label itself "Naturally Sparkling Mineral Water" on the bottles it imports into the United States.

"Sodium free" means less than 5 milligrams of sodium per serving (a serving is usually 360 ml). Even "very low sodium" and "low sodium" have specific meanings, respectively 35 milligrams or less and 140 milligrams or less per serving. "Natural" means that the water's mineral and trace element content have not been altered by processing. Evian, a water perhaps even slightly more chic than Perrier, is such a product—its mineralization in the bottle is exactly the same as when it came out of the ground. "Glacial" water has to come from a glacier; "artesian" water has to come from a source that flows above the water table. Kentwood Springs in Louisiana is the source of one well-known brand of artesian well water. "Well water" without the artesian label attached comes from a hole drilled into an aquifer.

Labeling laws don't cover pictures, so a bottler can put a picture on the label to suggest anything he wants. Pepsico, for example, the company that bottles a water called Aquafina, has a lovely design on the

label suggesting mountains, snow, and clear, cool water. But the stuff inside the bottle is the same as the water they make soft drinks from—processed municipal water supplied by whatever city the bottling plant is located in. A picture of a water-processing facility wouldn't do for the label—not at these prices. Coca-Cola sells a processed municipal water called Dasani, which they have labeled "purified water," but since they add minerals to it, the FDA has questioned whether it can be labeled that way. In any case, both Coke and Pepsi are processing, bottling, and selling water that they didn't dig for or pump out of the ground. It was delivered to them the same way it's delivered to you, right out of the tap, and it was perfectly drinkable before they got their hands on it.

Any of these waters, of course, can be polluted, either because the source itself is polluted or because the handling and bottling process introduces pollutants. This rarely happens in the United States. The CDC does report one outbreak of disease from bottled water in Saipan in the Northern Mariana Islands, a U.S. territory in which the rules of the EPA and FDA apply. Tap water in most parts of Saipan is too salty to drink, so two bottling plants treat municipal water (which comes from wells) by reverse osmosis filtration, an extremely efficient antibacterial filtering process. The exact source of the contamination was never established, but after drinking bottled water 11 people became ill from *Vibrio cholerae* serogroup O1, a bacterium that causes cholera, and four of them had to be hospitalized. (Cholera is generally a self-limiting disease and, provided that the symptoms of dehydration are treated, rarely has complications.)

Bottled waters vary in mineral content depending on the geology of the source. Waters from volcanic regions are often high in sodium, bicarbonate, and other minerals. Probably the most highly mineralized American bottled water is Mendocino Sparkling Mineral Water, which comes from a spring in the redwood forests north of San Francisco. It contains 240 milligrams of sodium per liter, plus plenty of calcium, magnesium, bicarbonates, and chlorides for a total of 1,500 milligrams of dissolved solids per liter. It is also naturally carbonated. Other bottlers, like Poland Spring (now owned by Nestlé, which also owns Perrier and Vittel), whose product contains only 46 milligrams of dissolved solids per liter, pride themselves on having almost nothing in their water.

Poland Spring water is filtered at the spring (or we should say springs, since Nestlé found several other sources of similar water in the

pine woods around South Poland, Maine) and passed through ultraviolet light for sterilization. Waters that are bottled at the source and not sterilized, distilled, or otherwise treated contain bacteria, and different sources contain different bacteria. Some believe that it is these benign bacteria that impart health-enhancing properties to bottled water.

And Another Parasite in Your Water

Giardia lamblia is the other common protozoan parasite that infests drinking water. During the 1990s, reports of giardiasis increased all over the country. In 1997–1998, the latest years for which the CDC has published statistics, New York State, including New York City, had the largest number of cases, 3,673. But before you conclude from this that New York City is a dirty place, you should know that the rate of infection in New York, 20.3 cases per 100,000 population, was less than half that of the rate in Vermont, 42.3 cases per 100,000 population. This probably won't stop bottled water distributors from using pictures of Vermont scenery on their labels to associate their products with clean, cool New England mountain streams, and it won't stop anyone from believing, contrary to observable fact, that New York City has dirty tap water. The highest rates of giardiasis, oddly, were among kids under five and people in their 30s. Most cases were reported during the summer and early fall, suggesting that swimming in polluted water has something to do with it (which we'll get to in the next chapter).

G. *lamblia*, also called *Giardia intestinalis*, is a one-celled parasite that lives in the intestines of both animals and people. It has a life cycle that involves at one stage a kind of cyst, an outer shell that protects the organism while it is outside the host's body. There are two trophozoites inside each cyst, and in this encapsulated form it can exist for an extended period in the environment, long enough to find a new host in which it can reproduce. Both forms are shed in stool, but the cysts are found mostly in nondiarrheal feces, the trophozoites in diarrhea. The trophozoite has two nuclei and four pairs of flagellae, and its appearance under a microscope suggests a cartoon extraterrestrial.

When a cyst gets into a person's intestines, the trophozoites "hatch" and begin multiplying by longitudinal fission—no sex is involved. They stay in the small intestines, either free-swimming or attached to the intestinal wall by a sucking disk. Then they pass into

the colon, where some of them become encysted, ready to be sent out into the world in feces to find another host. When they are swallowed by an unsuspecting human, this revolting process starts all over again.

While some people can be infected with *giardia* and not have any symptoms at all, others get the usual signs of intestinal illness—diarrhea, stomach pains, and cramps. In the worst cases, the diarrhea can lead to dehydration and can actually be fatal, and the symptoms can last up to six weeks or even longer. Everyone is at risk for this—child care workers, children in day care centers, travelers, hikers, campers, swimmers, anyone who drinks from a contaminated source or accidentally swallows contaminated water. Municipal water systems have on occasion become contaminated, causing wide outbreaks of disease. An outbreak occurred in New York State in June 1997 that affected 50 people. Its source was a surface water supply that was chlorinated but not filtered. Giardia is somewhat resistant to chlorine, but not nearly as resistant as cryptosporidium. A beaver was found near a valve box in the reservoir, but it was never established whether the beaver was infected.

Treatment usually isn't necessary, because the disease goes away by itself, but treating the illness may be advisable in young children who are persistently fatigued or losing weight. Medicine may also be required if several family members are ill, or if one member of the family is pregnant and therefore not able to take the most effective medicines. Metronidazole is the drug of choice, and there are several other prescription drugs available as well. *Giardia* is larger than cryptosporidium, so the same water filters that are effective against crypto will also work for *giardia*.

Two Bacterial Water Polluters

E. coli, which I mentioned at the beginning of this chapter, is usually transmitted in food, but it sometimes turns up in drinking water, too. In Wyoming in 1997, 157 people were infected with *E. coli* O157:H7 when a community water system supplied by a spring and two wells became contaminated. The water in this system is usually safely used untreated, but this time it was probably polluted by feces from wildlife near the spring. The same year in Illinois, three people got sick when they drank from an untreated well near a cattle pasture—the well was found to be contaminated with *E. coli*. A third outbreak that year

occurred at a trailer park in Washington State, this time associated with chlorinated ground water.

Shigella has occasionally found its way into water, too. There was an outbreak of *Shigella sonnei* infection in 1998 in Minnesota that made 83 people sick and sent four of them to the hospital. The source was a community water system that was supplying a local fair that had been accidentally cross-connected to the sewer system.

Home Waters

If you're worried about your own drinking water, you can have it tested. *E. coli*, shigella, cryptosporidium, and giardia don't have any taste and they're invisible, so you can't tell if they're in your water by looking or drinking. If your water comes from a municipal system, it is tested on a regular basis anyway, but you might want to know what effect your own plumbing has on your water and do a test on a sample directly out of your own tap. People buying a new house might want the water tested before they sign a contract of sale. If you are pregnant, planning to become pregnant, or have an infant less than six months old, testing well water for nitrate and coliform bacteria is important—pregnant women and infants should not drink water that has nitrate concentrations of higher than 44 parts per million. Those with wells that might be subject to contamination from nearby dumps, landfills, factories, dry cleaning operations, livestock pens, septic fields, mining operations, heavily salted roadways, or fuel storage tanks might also want to periodically test their water. If the water is staining your sink or your clothing, or if it looks cloudy or colored, these may also be reasons to test to see what's causing the problem.

Most state health departments will test water, either by collecting it themselves or having you bring a sample in to be tested. There are also private labs that will do the testing. You can't just say "test my water," though. You have to decide what you want the water tested for. There are hundreds of possible pollutants, and a corresponding number of tests to detect them. The most common (and cheapest) tests are for coliform bacteria and nitrates, but the tests can get quite elaborate and expensive when you get into testing for the many industrial and agricultural pollutants than can turn up in drinking water. In private labs, these tests can cost anywhere from $30 or less for the simplest tests up to $600 and more for the most elaborate.

Collection procedures and testing methods may vary from one lab to another, and it's important to collect the sample exactly as directed to get accurate results. Like any other area where there is money to be made, the water-testing field has attracted its share of questionable claims and dubious practitioners. Nonchemical water treatment devices are for sale that claim to eliminate various problems with magnetic impulses, electrostatic fields, or some other kind of force. People who sell worthless merchandise by making unscientific claims should be treated with the contempt they deserve. If you have doubts about a lab or a piece of equipment, the state health department can be a source of sound advice.

The latest report on surveillance for waterborne illness published by the CDC came out in May 2000 and covers the years 1997–1998. Thirteen states reported 17 outbreaks of illness associated with drinking water, mostly during the summer and fall months. Fifteen of the outbreaks occurred between May and October. No one knows the reason for five of the outbreaks. Of the other 12, copper poisoning caused two, and the remaining 10 were caused by parasites and bacteria: four by giardia, two by cryptosporidium, three by *E. coli* O157:H7, and one by *S. sonnei*. This seems like quite a few outbreaks, but as we will see in the next chapter, at least as many and often more people get sick by swimming in polluted water than by drinking it.

10

Fresh Air and Sunshine

OUTDOOR FUN WITH MICROBES

Suffolk County, New York, has beautiful beaches and beautiful people. Rich New Yorkers own elaborate summer houses there—in different neighborhoods from the local year-rounders who serve them, of course. Those who don't own can rent them—rentals at $20,000 a month and more are common. Some people commute to jobs in Manhattan from Suffolk. And some people who live there get very sick. One day an 11-year-old boy got a terrible fever that wouldn't go away. For five days he suffered with stiff muscles, stomachache, joint pain, and vomiting. His parents took him to the hospital on August 21, 1999. He had a temperature of 102°F, and several healing sores were visible on his skin. He was quite anemic—that is, the concentration of hemoglobin in his blood was way below normal. Something was destroying his white blood cells—his count was $2,800/ml^3$ when normal is between $4,500/ml^3$ and $13,500/ml^3$, a condition called leukopenia. His liver and spleen were swollen. At first they thought maybe he had either Lyme disease or babesiosis (a rare tick-borne disease), because the sores on his skin looked like insect bites, and both those diseases are spread by the same species of tick. But his blood tested negative for the organisms that cause those diseases. A chest X ray was normal, as were his serum electrolytes, a general blood test that can suggest many kinds of abnormal functioning. They asked his parents if he'd ever had a blood transfusion or an organ transplant, or if he'd traveled to foreign countries where he might have picked something up. He hadn't. So they did a blood smear—that is, took a blood sample,

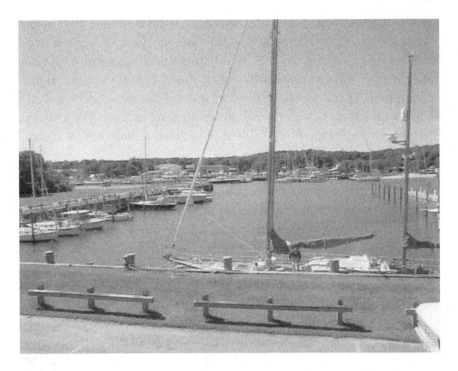

Figure 10 This is one of several marinas in Suffolk County, Long Island, New York. Hardly looks like the sort of place where you could catch malaria. But mosquitoes don't care how much money you make, and neither does *plasmodium*. *Suffolk County Government.*

smeared it on a slide, dyed it, and then examined it under a microscope. They found what they were looking for: *Plasmodium vivax*, one of the species of protozoa that causes malaria. Although they were never able to track one down, health department officials concluded that there were, somewhere on Long Island, anopheles mosquitoes transmitting malaria to humans.

Malaria

Rubbing on insect repellent and slapping at mosquitoes is a summer-time tradition. Just the smell of a spritz of Off! gives some people Proustian recollections of fun-filled days at summer camp. But until the 1920s, mosquitoes were a cause of deadly epidemics in the United States, and everyone now knows that they're back again carrying yet

another threat, West Nile virus, which we'll get to a little later in this chapter. Malaria, according to the latest genetic research, is a relatively new disease, too. It is now believed that malaria infections originated only about 8,000 years ago, when agricultural practices began to provide sufficient sunlit pools as breeding grounds for mosquitoes and sufficient population densities for humans to serve as hosts. Worldwide, there are an estimated 300 to 500 million cases of malaria, and about 2 million people a year die from it.

Some people are immune to malaria—in fact, the reason geneticists believe malaria only arose about 8,000 years ago is that that is when a gene variant arose that conferred immunity by causing changes in the genes that control red blood cells, which the malaria protozoan invades and destroys. Inheriting one copy of this gene from a parent confers immunity to malaria; inheriting two copies, one from each parent, causes sickle cell anemia, an often deadly blood disease.

The connection between mosquitoes and malaria was not always clear and was firmly established in 1897 by a researcher named Ronald Ross. He was awarded the 1902 Nobel Prize in Physiology or Medicine for his accomplishment. This led to using mosquito control as one weapon in the fight against malaria later in the century, but other factors helped contain the disease as well: population shifts to urban areas, improved drainage and housing, greater access to medical services, and the availability of quinine for treatment. Large-scale mosquito control with DDT did not begin until the 1940s. But malaria is still endemic in many parts of the world, including South America, and some scientists believe that it will begin to make a comeback in the United States over the next decade or two. The parasite is becoming more and more resistant to drugs, global warming may send infected mosquitoes northward, and there is no border control that can stop them.

Malaria is on the increase in the United States even without South American mosquitoes, likely because of increased immigration from malaria-endemic parts of the world. Malaria was probably introduced into the United States by European colonists and Africa slaves in the sixteenth and seventeenth centuries. It moved with people and eventually affected many areas of the country, with the exception of northern New England (for reasons we'll discuss in a moment). In 1934, when systematic reporting of malaria cases began, there were 125,556 cases. The number of cases dropped steadily through the late 1970s. Since

then, there has been a steady increase in rates, and presently about 1,800 cases a year are reported to the CDC, with between four and eight deaths caused annually by the illness. Most, but not all, of these cases are imported from other countries. Malaria is only carried by mosquitoes of the genus *Anopheles*, and there are several species within this genus. All 48 states in the contiguous United States have anopheles species capable of carrying malaria. So some of our plasmodium protozoa, like the one mentioned at the beginning of this chapter that made the Long Island boy sick, are probably home-grown. Between 1957 and 1994, there were 56 episodes of transmission, mostly in rural areas, but at least three outbreaks occurred in densely populated areas of New Jersey, New York City, and Houston, Texas.

Malaria can only be transmitted when a female mosquito bites a person infected with malaria, and the person has to have mature male and female stages of the parasite in his or her bloodstream. The male and female stages meet in the stomach of the mosquito where, in a rather complicated process of development and multiplication, they produce sporozoites, the infective form of the parasite. These sporozoites wind up in the salivary glands of the mosquito, whence they can be injected into the next person the mosquito bites. Thus epidemics begin.

There are four different species of *Plasmodium* that transmit slightly different kinds of malarial infection. *P. vivax* and *P. ovale* can persist in human liver cells, periodically sending parasites into the bloodstream and causing relapsing illness. *P. falciparum*, and *P. malariae* don't hide out in the liver, but if they aren't sufficiently treated, they can persist for months in the case of *P. falciparum*, and for years in that of *P. malariae*, causing relapsing clinical disease long after the initial infection.

In all four kinds of infection, the symptoms are recurring chills and fever, sweating, malaise, headache, and muscle pain—all symptoms that look a lot like, and are often misdiagnosed as, the flu. Rapid diagnosis followed by treatment with chloroquine, sulfanomide-pyrimethamine, or both together usually has excellent results, although some resistant strains are now developing in Southeast Asia, sub-Saharan Africa, and the Amazon basin. In cases where the concentration of parasites is high or where there is neurological dysfunction, intravenous quinidine gluconate is used, and this requires hospitalization in an intensive care unit. Otherwise, oral chloroquine is usually enough to wipe out the disease after a week of treatment. Therapy for vivax and ovale malaria,

Figure 11 A photograph of an anopheles mosquito having a meal of human blood. This insect invades various parts of the United States from time to time. While it is dining, it may also be injecting this person with plasmodium, the protozoan that causes malaria. *CDC. Photograph by Jim Gathany.*

because the parasites continue to emerge from the liver, requires treatment for about two weeks to make sure all the parasites have been killed. The most dangerous kind of malaria is falciparum, which can have a death rate as high as 20 percent if it isn't properly treated.

One of the things that limits contagion in the United States is that most mosquitoes don't live long enough to allow the plasmodium parasite to go through its life cycle. Below about 62°F, plasmodium can't survive to reproduce in mosquitoes. But warmer weather makes mosquitoes live longer and shortens the life cycle of plasmodium, and weather that is more humid and hotter than usual has been a factor in all recent outbreaks. This is probably why northern New England has been spared endemic malaria, while almost all other parts of the country have been infected. Suffolk County, New York, where two kids (one described previously) got malaria in the summer of 1999, is one of the most heavily mosquito-infested areas in the northeast, and 1999 was one of the warmest and driest summers in history. But just before the two boys got sick, there was heavy rainfall, which may have resulted in a large population of adult female mosquitoes. If you're a mosquito or a plasmodium protozoan, global warming may actually be good for you.

What are the odds of your getting malaria if you don't leave this country? Very low. Considering how many people spend summers outdoors and how many of those are bitten by mosquitoes, it's clear that 1,800 infections a year is a tiny percentage of those presumably at risk. If you use insect repellent, and if you wear long pants and long sleeves and don't go out at dawn and dusk when mosquitoes are feeding, your

chances of being infected are reduced even further. Fear of malaria should not in any way interrupt your plans for outdoor activity in the summer.

Lyme Disease

Malaria is not the only insect-borne disease now afflicting Americans. Lyme disease, named after the city in Connecticut where it was first identified, comes from the bite of a tick infected with a spirochetal bacterium. Two species of tick—*Ixodes scapularis* and *I. pacificus*—are the primary villains. *I. scapularis* lives in the East and Midwest; *I. pacificus*, as its name suggests, is found along the coast of California. But it may be that being worried sick about Lyme disease is more common than being sick with the bacterium that causes it.

You need three things in the environment for Lyme disease to occur: the bacterium, the tick, and mammals, usually deer and mice. Without these three elements, you can't have Lyme disease. Unfortunately, there are many parts of the country in which all three elements thrive together.

Ixodes ticks have a long and complicated life cycle that takes two years to complete. The adult ticks live on large animals like deer, where they feed on blood and then mate. This happens in the fall and early spring. By summer, the female ticks drop off the animal and lay eggs on the ground, where they hatch into larvae. These larvae attach themselves to birds and small mammals to get their blood meals. The following spring, they molt into nymphs. The nymph stage also feeds on small mammals and birds and then in the fall molts again, becoming an adult. It is at the nymph and larval stages that the tick becomes infected with *Borrelia burgdorferi*, usually by feeding on the white-footed mouse, and the bacterium continues to live in the insect after it becomes an adult. Then the adults feed on deer, starting the life cycle over again. When the tick bites a human, it inserts the bacterium into the person's bloodstream, where it causes Lyme disease.

These insects are tiny—hardly bigger than a pinhead in their larval and nymph stages, and only slightly larger as adults. They'll attach themselves anywhere, but they seem to prefer hairy and hidden areas—armpits, groin, scalp. The ticks—and in humans, it is usually the nymph stage that does the biting—are so small that they often go unnoticed,

Figure 12 *Ixodes scapularis,* the tick that carries the bacterium that causes Lyme disease. This is an adult female, and she looks big in the picture. Take a dime out of your pocket and look at the back. See the "ME" at the end of the word "DIME"? That's roughly the size of this insect, including the legs. The male is about half the size of the female, the nymph is about the size of the "E" in "DIME," and larva would cover about half the "I." *CDC. Photograph by Jim Gathany.*

and remain unnoticed during the two days of feeding usually required to transmit infection. The adults are easier to see, and therefore more likely to be removed before they cause infection. (Also, the adults live in cooler weather, when people are less likely to be outdoors where they can encounter them.) The larval stage, while it may be infected, usually doesn't carry the infection at the time it feeds, so it doesn't play a significant role in transmission.

Ticks don't jump, fly, or otherwise move through the air. And they don't live in trees. They crawl around in grass and shrubs, and you have to rub up against them to get them attached to you. If a tick makes it as far as your scalp, it crawled there from some other part of your body (unless you've had your head lying on the ground, which is of course possible). After they bite (and they do not insert their whole bodies, only their mouth parts) they spend about two days feeding before their meal is done. Mosquitoes, flies, and fleas don't carry Lyme disease. You can't catch it from another person no matter how close your contact. You can't get it from handling live or dead infected birds. You can't get Lyme disease directly from deer or domestic animals, you can't get it from the air, water, or food you consume, you can't get it from having sex, and although in theory you could get it from contact with blood or a blood transfusion, there's never been a documented case of that happening.

While the number of cases reported has gone up and down in the past few years, there is little question that the overall trend is upward. The geographic area affected expanded in the early years but has since

stabilized. In 1999, three states—Connecticut, New York, and Pennsylvania—accounted for 10,398 of 16,273 cases reported. There have been cases reported in all 48 contiguous states, but most of the cases are concentrated in the mid-Atlantic and north central regions in the East, and in a few counties of northern California.

The earliest sign of Lyme disease is a circular "bull's-eye" rash, red around the edges and white at the center, with the circumference growing over time. This is called "erythema migrans." There are also often nonspecific symptoms—headaches, muscle and joint pains, fatigue, malaise. The rash usually appears about 7 to 14 days after the infection, but it can appear as soon as three days or as many as 30 days after. As the disease progresses, the rash spreads, and disseminated infection may appear as a disease of the nervous system, the musculoskeletal system, or the heart. Neurological manifestations can include dysfunctions of the cranial nerves (including facial nerves) and spinal cord nerves. You can be infected without any symptoms—a blood test can determine this by detecting the antigens to *B. burgdorferi*. Having produced antigens to *B. burgdorferi* does not inevitably confer immunity to repeated infection.

Lyme disease accounts for 95 percent of all vector-borne disease (that is, requiring an animal carrier before you can be infected) in the United States. There have been more than 128,000 cases since reporting began in 1982, but the true figure may be larger because of underreporting.

The best way to avoid Lyme disease, obviously, is to avoid areas where there are ticks, especially in the spring and summer when the nymphs are feeding. But if your backyard is in, say, Connecticut or along the Pacific coast north of San Francisco, it may not be practical to follow this advice. Ticks like a moist, shaded environment, particularly overgrown grassy areas, leaf litter, and low-lying vegetation. So it's helpful to clean up these areas or just stay away from them. Wearing long-sleeved shirts and long pants tucked into socks or boot tops is also a good idea. And the application of insect repellents that contain DEET (n,n-diethyl-m toluamide) to clothes and skin also helps. A concentration of 35 percent DEET is effective; higher concentrations than that provide no additional protection. Permethrin, which kills ticks on contact, can be applied to clothes.

If a tick attaches itself to you, you have about 36 hours to get rid of it before it can cause Lyme infection, so checking for ticks and remov-

ing them when you find them is important. Before they bite—and they can be on you for hours before they settle down for a meal—they can simply be wiped or washed off. After they've attached themselves to the skin, fine-tipped tweezers are the best way to pull them off. Petroleum jelly, hot matches, nail polish, and other home remedies don't work. When you pull the tick off, it doesn't matter if the mouth parts remain in the skin—the bacteria are in the insect's stomach. The area should be cleansed with antiseptic when you're done. You only need medical treatment if the early signs of Lyme disease appear.

There are two vaccines for Lyme disease. One called LYMErix is manufactured by SmithKline Beecham Pharmaceuticals and can be used for people over the age of 15. The other is called ImuLyme, made by Pasteur Merieux Connaught, but as of early 2002 was not licensed for use in the United States. The Advisory Commission on Immunization Practices of the CDC recommends that people between 15 and 70 years old who live, work, or spend leisure time in areas of high or moderate risk and engage in activities that result in prolonged exposure to tick-infested habitats should get the vaccination. It can be considered for people whose exposure is less frequent, but it is not recommended for people who have only minimal exposure to tick-infested habitats.

Travelers to high-risk areas should consider having the vaccination, but should remember that they cannot achieve full protection until three doses have been given. Since doses must be given at 0, 1 month, and 12 months, it is a full year before the protection reaches its maximum. Also, vaccine administration (by intramuscular injection) should be timed so that the second and third doses are administered several weeks before the beginning of the *B. burgdorferi* transmission season, which usually begins in April. There aren't enough data to know whether boosters are necessary after two years. Considering everything, personal protection may be a better or at least equally effective approach for travelers.

Pregnant women shouldn't get the vaccination because its safety during pregnancy has not been established, and because acute Lyme disease during pregnancy responds well to antibiotic therapy without harming the fetus.

There are no data on the safety of the vaccine for people with impaired immune systems, those with musculoskeletal illnesses, or those with chronic joint or neurological illness, so they should not be vaccinated, nor should people who have treatment-resistant Lyme

arthritis or second- or third-degree atrioventricular block, a form of heart disease.

If you get Lyme disease, the latest research indicates that it can easily be stopped with a single dose of the antibiotic doxycycline. A study conducted in Westchester County, New York, divided 482 people who had found deer ticks on their bodies into two groups. One group got a single dose of doxycycline, administered in two pills. The other group got two placebo pills. The results showed that the single dose worked—only one of the people who got the antibiotic, 0.4 percent of the total, came down with Lyme disease. In the placebo group, 3 percent came down with the illness. This suggests two things: first, Lyme disease is hard to catch; and second, it's not hard to cure.

The antibiotic, however, can cause problems in itself—nausea, vomiting, and abdominal pain—and among those in the study who took it, there would have been little chance of getting Lyme disease even if they hadn't taken it. The latest recommendation is to watch the site of the bite to see if a rash develops. If it does, a few weeks of antibiotic treatment will cure the illness. Those who have symptoms after this are probably not suffering from Lyme disease, but from some other ailment.

In fact, most people who are infected get better without any treatment at all. A small percentage of infected people develop more serious symptoms—arthritis or heart disease—but even the majority of these people get better without treatment. Worrying about Lyme disease may cause more problems than Lyme disease itself. Dr. Leonard H. Sigal, a Lyme disease expert at the University of Medicine and Dentistry of New Jersey–Robert Wood Johnson Medical Center in New Brunswick, was quoted in the *New York Times*: "Lyme disease, although a problem, is not nearly as big a problem as most people think. The bigger epidemic is Lyme anxiety."

West Nile Virus

In late August 1999, the New York City Department of Health discovered an unusual outbreak of encephalitis (inflammation of the brain) concentrated among a group of elderly people living not far from La Guardia Airport in the northern part of Queens, one of New York City's five boroughs. The Department of Health, in cooperation with the Centers for Disease Control, began an epidemiological investiga-

tion, and within a week the CDC confirmed that the cause of the ill-
ness was an arbovirus, a disease caused by a virus carried by arthropods,
a phylum that includes ticks and lice, among other invertebrates. West
Nile virus, previously a disease confined to the Middle East, quickly
became the most talked-about disease in New York. By October 19,
1999, 31 cases of West Nile had been confirmed, and another 25 were
probable. There had been no onset of disease after August 22. The virus
is carried by mosquitoes and infects birds, mostly American crows, but
also ring-billed gulls, black-crowned night herons, mallards, robins,
red-tailed hawks, broad-winged hawks, and many others—76 different
species in all. Infected birds were found all over the New York City and
in an area of New York State stretching from Saratoga County near
Albany to Suffolk County at the tip of Long Island. Infected birds
and/or mosquitoes were found in New Jersey and Connecticut as well.

West Nile virus, first isolated in Uganda in 1937, had never before
appeared in the western hemisphere, although it is common in Africa,
eastern Europe, western Asia, and the Middle East. How it got here is
not known, although there are some theories. It may have been intro-
duced by an infected human, by a human who arrived along with
infected mosquitoes, or by a human- or storm-transported bird. It has
even been suggested that the introduction of the virus was the result of
an intentional criminal or terrorist act. The virus is a member of the
flavivirus family, closely related to the species that cause Japanese
encephalitis and St. Louis encephalitis. It is carried by several different
species of mosquitoes that like to feed on birds. The virus lives in a
mosquito for about 10 days and then migrates to the insect's salivary
glands, where it can be injected into an animal or human during the
mosquito's blood meal. The virus multiplies in the new host, causing
illness. Humans and animals other than birds are incidental hosts, not
essential to the life cycle of the virus.

Most people who get West Nile virus don't know it because they
are either completely asymptomatic or have very mild fever or
headaches before they quickly get better. A serological survey in
Queens conducted by the CDC on a sample of 677 people found 19
with undiagnosed West Nile infection—in other words, an astonishing
2.6 percent of the sample was infected without symptoms. This sug-
gests that thousands of people in Queens who had never been sick had
nevertheless been infected. The New York City Health Department
estimates that only somewhere between 1 in 130 and 1 in 300 people

who are infected show any symptoms. Some people get a rash; lymph glands can sometimes be swollen. When it gets serious, it can cause permanent neurological damage and be fatal, but less than 1 percent of those affected develop severe neurological disease. The worst form of encephalitis that the disease can cause leads to severe headache, high fever, stiff neck, muscle weakness, confusion, coma, and death. The elderly are overwhelmingly at highest risk. Among the group who develop severe neurological symptoms, death rates range from 3 percent to 15 percent.

In the summer of 2000, the number of birds and other animals with the disease grew substantially, with cases reported not only in birds, but also in chipmunks, squirrels, raccoons, bats, horses, rabbits, and other mammals. Twelve different species of mosquito were found infected with it. Moreover, the geographical distribution had widened to include cases detected either in humans, animals, or both in almost every county in New York, New Jersey, and Connecticut. Massachusetts, Rhode Island, New Hampshire, and Maryland also reported cases. Yet the number of human victims, for reasons that are poorly understood, decreased. Preventive spraying for mosquitoes certainly helped, but it is probably not the total explanation. In any case, final figures for 2000 issued in the spring of 2001 showed that only 18 people had confirmed cases of West Nile virus during the summer of 2000 (with a single death, that of an 82-year-old man), as opposed to the final figures for July to September 1999, when there were 62 cases and 7 deaths. On September 10, 2001, the New York City Health Department reported that three people had been infected during the 2001 season up to that time, and that 121 birds and 115 mosquito pools had tested positive for West Nile. In the meantime, the CDC reported that by July 2001 the geographic spread of the infection in animals had increased substantially up and down the East Coast, along the migrating paths of infected birds. By the middle of October 2001, more than 4,500 crows and other birds had been found infected, 125 other animals, mostly horses, had tested positive, and the virus had been found in nearly 700 mosquito pools. With all of that, though, there were only 31 human cases reported to the CDC.

When the final figures for 2001 were compiled, West Nile had appeared in various animals or mosquito pools in 359 counties of 27 states and the District of Columbia—a large increase from the 12 states that had reported the disease in 2000. The virus had also appeared in birds in southwestern Ontario. The total of human cases for the year was

66, with the dates of onset ranging from July 13 to December 7. Mosquitoes of the genus *Culex* are apparently the most common vectors of the illness in the United States.

The neurological symptoms of West Nile are almost impossible to distinguish clinically from those of other viral diseases such as herpes simplex or varicella, so a blood test is the only way to arrive at a certain diagnosis. But this doesn't mean that everyone who gets a mosquito bite should have his blood tested for West Nile virus. The tests (you need more than one to confirm the diagnosis) are expensive, laboratory capacity to do them is limited, and if there are no symptoms, they are unnecessary. Even people with very mild symptoms don't need to be tested—they'll get better by themselves. Another reason for limiting testing is that there isn't any vaccine or other medicine that works against West Nile virus, so the test results, although they may be important in epidemiological research, don't really matter to the individual—no matter what the test says, there's no treatment other than supportive care to provide symptomatic relief.

"Arboviruses"

The group of viruses that are transmitted by insects are called (and this is not a technical term) "arboviruses." The term is a coinage based on "ARthropod BOrne Virus." West Nile, transmitted by a mosquito, is one of these.

There are several other viruses transmitted by insects to humans, all of them quite rare in the United States. La Crosse encephalitis was discovered in 1963 in La Crosse, Wisconsin (another of those unfortunate cities that has lent its name to a disease). It has appeared since then in several Midwestern and Mid-Atlantic states. Unlike many viruses, it affects children more often than adults, and there are about 70 cases reported per year. It is transmitted by a mosquito called *Aedes triseriatus* and survives the winter in the mosquito's eggs, which grow into infected adults. Kids can get quite sick with this, and some of them wind up hospitalized for central nervous system infections. The virus can also cause neurological problems that last for several years. The illness can kill, but it rarely does: the fatality rate is under 1 percent.

St. Louis encephalitis is caused by a flavivirus closely related to West Nile. It's transmitted by several different species of mosquito,

depending on the region of the country, and there are an average of 128 cases per year. The disease occurs all over the country except in northern New England. The virus is carried by birds, but unlike West Nile virus, it doesn't make the birds sick. Normally the virus travels back and forth between birds and mosquitoes—people and other animals are "dead-end" hosts, not part of the normal life cycle of the virus. Most infections result in either mild symptoms or no symptoms at all, but the cases that have symptoms are overwhelmingly central nervous system problems, including neck stiffness, stupor, disorientation, tremors, convulsions, paralysis, and coma. The illness in its serious form affects mainly the elderly. You can only get St. Louis encephalitis from the mosquito—it isn't transmitted person-to-person or by contact with any other animal. There's no vaccine, and no treatment except supportive therapy.

Two other arboviruses that are carried by mosquitoes and cause encephalitis are western equine encephalitis (WEE) and eastern equine encephalitis (EEE), caused by different types of a virus of the *Togaviridae* family, genus *Alphavirus*. These infections are rare, but they can be deadly. Even though it is very uncommon, EEE is the most severe of the arboviral illnesses—in 1996–1997 there were 19 cases of EEE, and five of them were fatal. It is thought that the virus may be amplified in emus, large flightless birds that are raised commercially for meat and other purposes, making humans more susceptible to the illness, even though there is no evidence that an emu has ever directly infected a human. There are more cases of EEE in Florida than in all other states combined.

EEE and WEE, as their names suggest, both infect horses. The CDC doesn't track the horse version of the disease, but some state health departments do, and there are more than a hundred cases each year, with 23 states reporting. Humans can't get the disease from a horse, only from a mosquito. There is neither a drug to treat nor a vaccine to prevent either EEE or WEE.

The man from Kennebec County was a real Down Easter who hadn't left Maine since he was 45—a quarter-century earlier. He'd been spending most of his time doing yard work and lying on the ground repairing the hull of a boat, when in June 2001 he got really sick: muscle weakness, somnolence, diarrhea, and a fever of 104.7°F. His kidneys weren't working right, either, and he was anemic. Shortly after hospi-

talization, the left side of his body became paralyzed, and his mental state became completely confused. He'd had a stroke, clearly, but no one could figure out why. His cerebrospinal fluid contained 10 times as many white blood cells as normal, suggesting a severe infection of some sort. A serum test found antibodies to a very unusual virus, the one that causes Powassan encephalitis. He was hospitalized for three weeks until his acute symptoms subsided and was then discharged to a rehabilitation facility. Three months later, in September 2001, he was still paralyzed on his left side.

Powassan encephalitis has been known since 1958, but in the 40 years that followed its discovery, there were only 27 known cases of the illness. Then, between September 1999 and June 2001, there were four new cases reported in Maine and Vermont, including the one described above. Each of the patients suffered symptoms similar to the man from Kennebec. The Powassan virus is carried by four different species of tick in North America, and 38 mammal species have been found to be infected, mostly woodchucks. Unlike the tick that carries Lyme disease, the ones that carry Powassan virus rarely look for hosts walking through the brush. Instead, they are found near the nests and burrows of medium-sized mammals. Although two of the victims didn't recall any tick bites, they all lived where the ticks flourish, and the insects are easily missed. Powassan is rare but deadly: there is a case-fatality rate between 10 percent and 15 percent, and long-lasting neurological deficits are common. There's no vaccine and no treatment, so the only protection is prevention. The CDC believes that because of lack of awareness and the need for specialized laboratory tests to confirm a diagnosis, the illness may be more common than previously thought.

The Trouble with Swimming

Being bitten by infected ticks and mosquitoes, alas, is not the only germ danger you face while enjoying the great outdoors. In fact, just going for a swim can make you sick, even if you never get near an insect. *Cryptosporidium*, *Giardia*, and *Shigella* are out to get you, and they're lurking where you swim, summer or winter, ocean, lake, river, or pool. Not to mention the Jacuzzi at the spa, the decorative fountain the toddlers wade in, and the water park where the kids had such a great time. Sometimes it can take several weeks after swimming in polluted water for a

diarrheal illness to begin, and therefore people don't often connect the swimming with the illness. Many people don't even realize that swimming in a chlorinated pool can make you ill. In addition, diarrhea is fairly common, and most people wouldn't bother reporting diarrhea even to their own doctor, much less to the health department. So the CDC's statistics are almost certainly an underestimate of the actual number of outbreaks.

Swimming is the second most popular exercise in the United States, and there are about 400 million pool visits every year. In 2000, there were five outbreaks of swimming pool–linked cryptosporidiosis reported to the CDC. This made several thousand people sick, as such outbreaks do every year. Cryptosporidium isn't a bacterium and it isn't a virus. It's a protozoan parasite.

As you know from chapter 9, crypto can get into water supplies, and when people drink water infested with crypto, they get bad diarrhea. People usually don't die from the illness (there are some deaths among the very old, the very young, and the immunocompromised), but it's still no fun. You might think that a well-chlorinated swimming pool would make such an infection impossible, but you would be wrong. In fact, crypto is highly resistant to chlorination, and even the best-maintained swimming pools can be full of it. If you go swimming in one of these pools, and if the water gets into your mouth, you can be infected, too.

How does crypto get into a swimming pool in the first place? People's rear ends, to put it delicately, are typically not very clean places. And when people go swimming, their rear ends get washed off in the water. If a person goes swimming with diarrhea and has a "fecal accident," a large number of cryptosporidium can be released into the water, where people have a good chance of swallowing some of them. It isn't easy to know when someone has had one of these accidents, and crypto can survive for several days in chlorinated water, so you can't really know when you're entering a pool that is infected.

Crypto has a clever way of protecting itself in the outside world until it can find a new home inside another host. The parasite is excreted from an infected person in an oocyst, a kind of protective shell. It doesn't "hatch" until it is ingested by another person, in whose gastrointestinal tract the oocyst breaks open and the sporozoites inside attach themselves to the epithelial cells. Parasitizing these cells, the protozoan undergoes asexual multiplication by cell division, and then sexual multiplication in which female macrogamonts are fertilized by male

microgametes. Next oocysts are generated, which contain the new young sporozoites. The sporozoites, encased in their oocyst armor, wait in the host's feces until they are excreted to the outside world, where they can seek a new host. Chlorination in the concentrations used in swimming pools is no barrier to crypto's continuing reproduction.

One "fecal accident" can pollute an entire swimming pool. And very few organisms are required to infect a human. If you swallow just one or two mouthfuls of water from that polluted pool, chlorinated or not, you can get a case of cryptosporidiosis.

Where else is crypto found? Since crypto is resistant even to chlorinated water, you can easily imagine that unchlorinated water is an even safer haven for the bug. If you swim in lakes or rivers, you can still get crypto. Bathing in Jacuzzis, decorative fountains, or hot tubs is also a source of infection. And since crypto is resistant to salt water as well, you can even get it by swimming in polluted ocean water. All you have to do is swallow some of the water.

The disease itself, cryptosporidiosis, is very unpleasant, but it usually goes away by itself without treatment. It can last for up to two weeks, with symptoms disappearing and then returning during that period, and the diarrhea, stomach cramps, and fever that can be associated with it can be debilitating. There's no medicine that cures it, and no treatment other than drinking plenty of fluids. Usually the disease is so mild that people get crypto without even knowing it, it goes away, and they are none the worse for wear. Unfortunately, crypto has yet another neat survival trick: it can stay in you for up to two months after the illness has disappeared, and during this period, you can pass the disease on to others. Since you never knew you had it in the first place, you don't know you're still infectious. This is great for the parasite, but not so good for people you go swimming with.

Can you test the water for crypto before you swim in it? Well, there are tests, but it is very unlikely that you are equipped to carry them out. For one thing, the tests require many hours of analysis by a specially trained microbiologist. For another, you'll have to supply that microbiologist with somewhere between 100 and 400 gallons of water to do the tests. And of course testing the water one day doesn't mean that it can't be infected the next. Public municipal water systems do periodically test for crypto, but a home test for your own swimming pool or hot tub is simply not realistic. Your best bet is to prevent crypto from getting into your swimming water in the first place,

and helping to protect others by not entering public swimming areas if you have diarrhea, and not allowing your kids to swim if they have diarrhea. Infants who are not toilet trained should not bathe or swim with other people.

There are two ways the reproductive cycle of crypto can be stopped: crypto can't survive boiling for more than a minute, and the parasite is big enough to be filtered out of water, provided you have the right kind of filter, but conventional swimming pool filters are not the right kind (see chapter 9 for details). So boiling and filtering can prevent you from getting crypto from drinking water, but they obviously can't help at all when you go swimming.

It Only Takes One to Make You Sick

Antonie van Leeuwenhoek loved to look at stuff through his microscope. One day in 1681, he was closely examining a sample of his own feces and described a parasite that was probably the one we now know as *Giardia lamblia*. It got its name, however, in 1859 from a biologist named Lambl. It is the most frequent cause of nonbacterial diarrhea in North America, and you can get it by swallowing polluted water where you swim. Like *Cryptosporidium*, *Giardia* is protected when it's not inside a host by a cyst, though it is not quite so resistant to chlorine as crypto is. Conveniently (for the parasite, that is), human stomach acid is perfect for dissolving the cyst and allowing the trophozoite to "hatch," after which it attaches itself to the wall of the intestine by using a sucker on its ventral surface. Then it reproduces asexually by binary fission. Some researchers think that the parasite produces a toxin that causes the illness; there is another theory that the organism somehow obstructs the normal absorption that goes on in the intestines. There are various strains that can be distinguished by analysis of their DNA, but there doesn't seem to be any consistent relationship between the kind of strain you get and the severity of your disease. All you need is one cyst to make you sick—unlike bacteria, of which you can consume thousands before you get ill. Since kids get sick with *Giardia* more often than adults, it's likely that infection confers immunity, but no one is quite sure. In addition to getting *Giardia* infection when swimming, you can also get it from food and from contact with feces, such as while diapering babies. (The incidence in day care centers is appalling—

sometimes as high as 50%.) The disease is common in homosexual men, suggesting that sexual transmission is possible as well.

Swimming in polluted water is probably the most common way of getting this infection. *Giardia* is large enough to be filtered out of water systems, but this doesn't prevent fecal contamination of recreational water. Metronidozole is the best treatment, but there is a small minority of sufferers, less than 4 percent, who remain symptomatic longer than two weeks and whose infections are quite difficult to treat. Sometimes the chronic disease has serious consequences—specifically, a syndrome in which food can't be absorbed properly, leading to severe weight loss. This chronic form, though rare, is more common in adults than children. *Giardia* infection can also be completely asymptomatic. In fact, about 7 percent of stool samples submitted for microscopic examination contain the cysts of the parasite, which makes it the most commonly observed flagellate in the intestinal tract.

The symptoms of *Giardia* infection are no fun. It usually begins with nausea and lack of appetite. Sometimes there is a slight fever and chills. Later, explosive, watery, foul-smelling diarrhea begins, accompanied by the passing of foul gas and, often, belching that has a bad taste. Cramps are common, too. Still, the only way to know for sure that it's giardiasis is by seeing it in a stool sample, just as van Leeuwenhoek did more than three centuries ago.

The same rules about prevention of crypto in recreational water apply to *Giardia*. Since it's transmitted in feces, the best way to prevent it is to keep feces out of the swimming water. Don't swim with diarrhea, and don't let your kids do so, either. Even if they don't have diarrhea, washing little kids' behinds before they go in is a good idea. While you're at it, wash your own, too.

The CDC issues recommendations about what to do when there are fecal accidents in swimming pools (they call these "recreational water venues" to cover pools, water parks, wading pools, fountains, Jacuzzis, hot tubs, and any other place where people bathe together in chlorinated water). There are detailed procedures for removing waste matter, raising the chlorination levels, backwashing filters and replacing filter media, keeping people out of the pool for varying periods of time depending on the incident (whether it is formed stool or diarrhea), measuring the results of the preventive measures taken, and maintaining a log of incidents that includes date and time of the event, formed stool or diarrhea, free available chlorine concentrations, pH levels, and

so on. "Recreational water venues" that conscientiously follow these procedures are likely to be the safest places to swim.

What happens if the pool doesn't follow the best maintenance procedures? Usually nothing, but then occasionally something very bad occurs. In August 1999, the Volusia County (Florida) Health Department received reports of three kids who had been infected with *Shigella sonnei. S. sonnei* is one member of a group of bacteria that cause gastrointestinal illness. The disease, shigellosis, has the usual revolting symptoms—diarrhea, sometimes bloody, stomach cramps, and fever. Young kids are especially at risk for it, and sometimes their disease is so severe that it requires hospitalization for intravenous rehydration. In kids under two, it can cause seizures. There are about 8,000 cases reported every year, and since you can have it without symptoms or with very mild symptoms, there are probably many more infections than that. Although it can be treated with various antibiotics, it is starting to develop resistant strains; in any case, otherwise healthy people recover from mild cases without treatment. The use of antibiotics, therefore, should be limited to the most severe cases.

The only thing the three Florida children seemed to have in common is that they'd all been in an "interactive water fountain" that had opened on August 7 in a park near an ocean beach. (An interactive water fountain is one intended for playing in, as opposed to looking at.) The health department got very curious about this interactive fountain, so they interviewed a sample of 86 people who'd been playing in it. Thirty-eight of them, or 44 percent, had gotten sick, most of them with diarrhea, abdominal cramps, fever, and vomiting. Thirteen percent of them had bloody diarrhea. The people who got sick were an average age of eight years old, younger than the group that stayed well.

Then the health department took a look at the fountain itself. It worked by recirculating the water through a hypochlorite tablet chlorinization system with no filters. The problem with a tablet system like this is that the tablets wear out after about a week and are no longer effective. The tablets hadn't been changed since the park opened three weeks earlier, and no one had monitored the chlorine levels of the water. Diaper-aged kids loved this fountain, and they would routinely stand directly over the nozzles, where the spray did a nice job of washing their behinds. In other words, this fountain was a nearly perfect place for culturing and spreading germs, which it did with considerable

efficiency. All you had to do was swallow a couple of mouthfuls of the water, and all but two of the people who got sick did exactly that.

While giardia, crypto, and shigella are the most common problems in swimming water, there are others, much rarer. One deserves mention if only because its life cycle is so bizarre and its effect on humans so deadly. There have been few cases of infection with a protozoan called *Naegleria fowleri*, but the ones that happen are truly horrifying. Like giardia and crypto, *Naegleria* protects itself with a cyst when necessary, and then breaks out of the cyst to reproduce inside a host. It feeds on bacteria such as *E. coli*, a habit as adaptive as it is revolting. The protozoan enters the body through the nasal passages, typically when the person is swimming or bathing in hot springs or hot baths. Inside the nasal passages, the protozoan comes out of its cyst and migrates to the olfactory nerves, eventually invading the brain. After about a week of incubation, the symptoms appear abruptly: rapidly progressive headaches, fever, nausea, vomiting, inflammation of the pharynx, lethargy, confusion, stiff neck, convulsions, and coma. Unless the illness is treated (amphotericin B has been used successfully), death can occur as fast as 24 hours after the first symptoms appear, though sometimes it takes as long as two weeks. The way to diagnose the illness is do a lumbar puncture and examine the cerebrospinal fluid for evidence of the parasite—but because the disease is so rapidly fatal, there's often little or no time for this.

Although this distressing disease is rare, many people have antibodies to the parasite, so exposure and asymptomatic infection are probably fairly common. But we're only telling you about this one to give you something to have bad dreams about. The fact is, you're extremely unlikely ever to suffer from it.

Rabies

Everyone's afraid of rabies, but almost no one in the United States ever gets it. Even a hundred years ago, it was very rare—about 100 people a year got it in the early twentieth century—and today, thanks to widespread vaccination of domestic animals, the average yearly number of human cases is 2 or 3; there were only 32 cases from 1990 until the end of 2000. In 1999 there were no cases of human rabies at all—zero. Yet

however rare it is, it's 100 percent fatal if untreated, which is reason enough to be afraid of it. In the fall of 2000, five men in five different states were diagnosed with rabies, and every one of them died, the first human rabies diagnoses since December 1998. One of the men had just arrived in New York from Ghana and had been infected in his native country by a puppy. Rabies is much more common in developing countries than in the industrial world, and canine rabies is prevalent in Africa, Asia, and Latin America. Of the 50,000 rabies deaths each year, more than 95 percent are reported in areas where canine rabies is common. But the other four were not bitten by rabid dogs. Instead, each of them had had some kind of contact with bats. None of the five lived more than two weeks after being hospitalized.

Between 1990 and December 2000, 24 of the 32 cases of rabies reported to the CDC were attributable to bat-associated variants of the virus, even though it couldn't always be firmly established that a bat bite was the cause of the illness. One problem is that bats are quite small, and their bite often seems a minor injury unworthy of medical attention. Moreover, some people, especially children or the disabled, can be bitten by bats without reporting it or even knowing that they have been bitten.

Rabies can be carried by all species of mammals, and it's just as fatal to them as it is to humans. It causes encephalitis, swelling of the brain. Different species carry different variants of the virus, and in addition to bats, raccoons, skunks, foxes, and coyotes are major reservoirs of the virus in the United States.

Although there are other ways of getting the virus, they are very rare in comparison to the most common way: by the bite of an infected animal. When a raccoon, for example, is bitten by a rabid animal, the virus enters with the infected saliva, spreads through the nerve to the spinal cord, and then incubates for one to three months. During this period, the animal shows no sign of illness. As the virus moves to the brain, it multiplies and passes to the salivary glands. At this point the animal is sick, and the sickness is brief: within a week, the raccoon is dead.

In humans, the first sign of rabies infection (as with so many other illnesses) is flulike symptoms: malaise, headache, and fever lasting for several days. Soon brain function is affected, causing anxiety, confusion, agitation, hallucinations, and insomnia. By the time the clinical symptoms have appeared, it's too late: there are exactly six documented

cases of people surviving after symptoms have appeared, and all of them had had some kind of pre- or postexposure prophylaxis.

The only way to avoid dying after being infected with rabies is to get immediate prophylactic treatment before symptoms appear. There are several tests required if you want to make sure you haven't been exposed—no single test will rule out the possibility. These tests include serum, spinal fluid, skin biopsy, and saliva examinations to test for the presence of antibodies to the virus. It's easier to figure out if an animal is rabid—you kill it and test its brain tissue, a procedure that can be carried out within a few hours. Knowing if the animal that bit you was rabid is obviously very important. If it wasn't, you can relax. If it was, you need prophylaxis immediately.

There is a very effective rabies vaccine that was developed about 20 years ago. It works when administered before exposure, and works equally well when administered just after exposure. About 58,000 people in the United States are treated every year with the vaccine. Most people don't need preexposure vaccination—that's reserved for high-risk groups such as veterinarians, animal handlers, and people who work with certain laboratory animals. Anyone who has frequent contact with potentially rabid raccoons, bats, skunks, cats, dogs, or other mammals should also be vaccinated, and if you're traveling to areas of the world where rabies is endemic, you should consider getting the vaccine.

If you're exposed to an animal that is possibly rabid, then you need the vaccine as soon as possible, as well as one dose of immune globulin. Postexposure, the rabies vaccine is given in five shots over a 28-day period, and the immune globulin is administered along with the first shot. The vaccine, contrary to what you may have heard, is relatively painless and given in the arm, just like a flu shot. It has never, not even once, failed to work when given promptly. "Promptly" does not mean that you have to panic—the CDC calls it a "medical urgency," not a "medical emergency." The incubation period of the virus—that is, the time between the initial contact and the appearance of symptoms—is very variable and can be anywhere between a few days and several months. But there is time to consult with your physician and with public health officials regarding the need for the vaccine.

If you are bitten, the health care worker will want some facts, including where the incident happened, what kind of animal bit you, whether the exposure was provoked or unprovoked, whether the animal has been

vaccinated, and whether the animal can be or already has been captured so that it can be tested. The most important thing is to seek care promptly. As we noted, rabies is, for all practical purposes, 100 percent fatal if you don't treat it, and 100 percent preventable if you do.

You can't get rabies from another human by casual contact such as touching or contact with urine, blood, or feces of an infected person. Theoretically, you could get it from being bitten by a human with rabies, but there has never been such a case reported. Nor does contact with a person receiving postexposure prophylaxis mean you've been exposed to rabies. The only human-to-human transmission ever documented happened among eight people receiving corneal transplants in five different countries (two in Thailand, two in India, two in Iran, one in the United States, and one in France). New guidelines for the acceptance of corneas for transplant have reduced this tiny risk even further.

Why bats? There seems no reason to believe that bats are any more susceptible to rabies infection than many other mammals, but it may be that close human contact with them is more common than with any of the other usual wild reservoirs of the virus: raccoons, foxes, skunks, and coyotes. Unlike these other mammals, bats often get into houses, where they nest and breed. Other small wild mammals that live close to people or in their houses—rodents like mice, squirrels, and chipmunks, for example—have never been known to cause rabies among humans, even though they can theoretically be infected with the virus. The rodent that is most commonly infected is the woodchuck (sometimes called a groundhog), but most people only see them from a speeding car, sometimes alive, and probably just as often squashed beyond recognition in the middle of the road. So that leaves bats, which most people find a rather unappealing creature to begin with, for at least one good reason and several bad ones.

Although there is a species of bat, *Desmodus rotundus*, that feeds on the blood of vertebrates, including humans, these bats are tropical creatures that do not live in North America. North American bats do not suck your blood, they're not blind, they're not birds, and they're not rodents. There are several different species that are common in North America, including the red bat, the hoary bat, the Mexican free-tailed bat, the big brown bat, the little brown bat, and the silver-haired bat. This last species is the one most commonly associated with human rabies. Bats are mammals, and their diet consists mainly of insects. Eat-

Figure 13 The big brown bat, *Eptescius fureus*. It looks like a killer, but it isn't. Except when it has rabies, which it sometimes, though rarely, does. *CDC. Photograph by Dr. Sikes.*

ing insects, especially agricultural pests, makes them an essential part of the ecosystem everywhere in the world from rain forests to deserts. Living safely with them is important to everyone. Although the only sure way of knowing if a bat is rabid is by a laboratory test, any bat found where bats are not normally seen—in the middle of the lawn during the day, for example—or a bat that appears unable to fly should be considered suspect. If you wake up and find a bat in your room, you should seek medical attention even if you can't find a bite on your body. You can't get rabies by seeing a bat in the attic or in a cave or at a distance, and you can't even get it from contact with bat guano or urine. Touching a bat is a bad idea, but you can't get rabies by touching its fur. Only getting the saliva into your body thorough a wound, your eyes, your nose, or your mouth will do it.

Well, enough of the great outdoors. Let's go home, take a shower, get dressed, and go out to a restaurant for a delicious dinner. Let's relax a little. But not too much . . .

11

Paint the Town Red

GERMS IN PUBLIC PLACES

You're relaxed, sipping your wine, waiting for your appetizer, looking forward to a nice quiet dinner: broiled lamb chops provençal, your favorite. The people at the next table have just left, and their dirty dishes are being collected by the busboy, who was rubbing his cheek—or was it his nose?—on his way over to the table. He's grabbing the knives, forks, plates, and glasses, piling them up along his arm for transport back to the kitchen. His fingers slide across the surface of the soiled plates as he picks them up. You stare down at the tablecloth, thinking. You suddenly remember that the exterminator is coming tomorrow, and you have to leave the keys so the super can let him in. You wonder if they have exterminators come to this restaurant. To the kitchen. The kitchen where your dinner is being prepared. What would they be exterminating in there? Do they do the bathrooms, too? "Employees Must Wash Hands Before Leaving the Bathroom." Well, yes, but it doesn't say they have to wash their hands after they pick their noses, does it? And who pays attention to signs anyway? How about the chef? Did he wash his hands? He's French—probably doesn't even speak English, much less read signs in the bathroom. He uses his hands to arrange the food on the plate, doesn't he? Otherwise it would never look so pretty, so perfectly balanced, those lamb chops piled precisely on top of the summer squash, the white wine sauce with minced shallots dribbled over the top, the plate sprinkled with ground parsley. That looks like hands-on work, doesn't it? Does anyone else help him push food around your plate? And does his nose itch, too? What's in that

busboy's nose that he just casually transferred to his hands? Does he get to touch any food back there? After he touches those dirty plates?

Still hungry?

Restaurants, when you start to think about it (which you probably shouldn't just before dinner) are a good place to pick up something nasty. And germs don't really care how fancy or expensive the restaurant is—they just want a nice place to live and make more germs. When you know that Nobu, the pricey, superchic Japanese restaurant in the SoHo neighborhood of Manhattan, frequented by actors like Robert De Niro and supermodels so skinny that they're clearly not there for the food, was cited by the New York City Health Department in February 2000 for having "vermin or other live animals present in food storage, preparation or service area," you know you can get in some kind of trouble wherever you pick up a fork (or a chopstick).

"Vermin" sounds pretty bad. The New York City Health Department leaves nothing to the imagination. Vermin, they specify, is an "insect or small animal that is destructive, annoying, or harmful to health. Includes cockroaches, flies, and rats." Nice. Most people who are told that there are "cockroaches, flies, or rats" in the area of the restaurant where the food is prepared will probably want to find another restaurant—even if Al Pacino is sitting at the next table. What does the presence of insects and rodents where food is prepared portend for your continued good health? Let's take a look.

Your Insect Roommate

There are dozens of different species of cockroach, and the one probably most familiar in restaurants is the German cockroach. Many people believe that cockroach infestation is one of the burdens of living in a crowded city, but in fact the insects can and do live in the wide open spaces of rural areas just as happily. Maybe a cockroach is just more noticeable when it marches across your menu. Cockroaches are actually pretty clean—at least by their own standards. Their bodies are covered with sensory bristles that must be kept scrubbed and neat in order to function properly, so cockroaches expend a lot of effort in grooming themselves and are actually quite scrupulous about their own hygiene. Unfortunately, they're not so scrupulous about yours. They'll eat almost anything—rotting vegetable matter, houseplants, vegeta-

Figure 14 The German cockroach, in various stages of development. Feel free to express your revulsion, but don't try to pretend you've never seen one before. No one will believe you. *Department of Entomology, University of Nebraska-Lincoln.*

bles, meat, a crumb of bread, a dead mouse—and they leave droppings all over the food, which they then step in and transfer to the next pile of food they attack. The stuff in their droppings is really gross: all sorts of pathogens that can cause diarrhea, dysentery, typhoid, lung ailments, urinary tract infections, and other illnesses.

One study found that the German cockroach, America's favorite, carries a great variety of bacteria, and a little more than half of the kinds they carry are disease-causing. Among the most common pathogens found on or in roaches were *Klebsiella oxytoca, Klebsiella pneumoniae,* and *Enterobacter cloacae.* These are gram-negative bacteria of the family that includes *Salmonella* (famous in food poisoning cases), *Shigella* (the main cause of human dysentery), and *Yersinia* (one species of which causes the plague). *Klebsiella* can cause infections in the lungs and the urinary tract that can be fatal under certain circumstances, and it is resistant to many antibiotics. *Enterobacter,* which is almost exclusively a nosocomial infection (that is, transmitted in hospitals), also causes disease in various parts of the body. Cockroaches carry the bacteria on their exterior cuticle, contact with which alone is enough to spread them, and in their guts. Their feces are full of them.

Allergy to cockroaches is a risk factor for asthma. Asthma is a serious health problem—more than 5 percent of the U.S. population has asthma, and the number is growing. Of course, cockroaches are only one of many risk factors for asthma sufferers in the indoor environment—household dust mites, fungi, pollen, mice, rats, and dander from pet birds, cats, and dogs also contribute. Still, there is no doubt that cockroaches are in general not good for asthma sufferers.

That's the bad news. The good news is that despite their ubiquity, their unsanitary toilet habits, and their ability to carry and transmit all

sorts of pathogenic microbes, cockroaches have never been implicated in any major disease outbreaks in the United States. They're ugly. They're disgusting. They're a threat to proper digestion, because just thinking about them can turn your stomach. They certainly carry disease-causing microbes, and most people would rather not see them in a restaurant (or anywhere else, for that matter). But a serious health threat? Maybe in some circumstances, but apart from their association with asthma, probably not much of one.

Mice and Rats

Some people love mice and rats and keep them as pets. The white ones with the pink eyes are nice, very soft and furry, and many people love them as much as they love their dogs. And they may be man's best friend in another sense as well: they are by far the most common experimental animal in medical research. There are hundreds of specially bred lines of mice and rats used for research in various fields of medicine, biology, physiology, psychology, and so on. There are mice that easily get skin and lung tumors but are resistant to breast tumors, rats that commonly have vaginal prolapse, rats that are aggressive, rats that are passive, mice that have a high incidence of testicular birth defects, mice that have abnormal kidney function, skinny rats, fat rats, mice that have a high incidence of congenital cleft palate when their mothers are injected with cortisone, mice that tend to chew each other's whiskers off, mice that are immune to tetanus, rats that like to run in wheels and rats that don't, rats that get asthma, rats that are highly susceptible to malignant tumors of the central nervous system, mice that are resistant to *Herpes simplex*, mice that are born deaf, and hundreds and hundreds of other genetic variations. It almost seems that if there's a behavior, a disorder, or a disease you are interested in, there's a rat or mouse strain genetically bred to manifest it.

Mice are often used in vaccine development, and until the middle of the twentieth century, they were important in the production of vaccines as well. Today, however, vaccines are produced by tissue culture without the involvement of live laboratory animals. Bacterial vaccines still require serum or blood for culture, and mice are often used as part of the production process for these medicines. Animals, and particularly mice and rats, are widely used in testing for toxicity, tumori-

genicity (potential to cause tumors), and potency of many drugs, and for testing the purity and safety of vaccine batches.

And then there are those mice and rats that carry diseases without human intervention—pests that have tormented humankind throughout history, and probably even before history was recorded. These are the rats that were a vector of the Black Death in medieval Europe. These are the mice that carry hantavirus in the American West. These are the animals that figure in the life cycle of the bacterium that causes Lyme disease. And these are the rodents that the New York City Health Department vividly characterizes, along with cockroaches and flies, as "vermin" when they turn up in a restaurant. Various species of rats and mice live only in close proximity with humans, and in most parts of the world, where humans live, these rodents live with them.

If you read the last chapter, you know that mice and rats don't carry rabies (although, like many other mammals, they theoretically could). But there are plenty of diseases they do carry, and a number of them can be transmitted to humans under the right circumstances. In fact, mice are carriers of some of the most deadly viruses known, the various viruses that cause viral hemorrhagic fever, or VHF. While these viruses are not common in the United States, they do occasionally appear, and when they do, it's not pretty. Their symptoms include bleeding from the mouth, ears, eyes, and internal organs; kidney failure; coma; delirium; seizures; and death.

There is only one place in the United State where four states have a common border: the place where Utah, New Mexico, Arizona, and Colorado all come together, sometimes called the Four Corners. In 1993, a doctor working for the Indian Health Service in Gallup, New Mexico, noted a cluster of five deaths in the Four Corners area and told the health department about it. No one knew what the problem was, so they organized a conference of doctors and health officials to try to figure it out. Even as they were organizing the conference, more reports of death were coming in. People were coming down with fever, chills, and muscle pain, followed by cough, thrombocytopenia, and elevated serum levels of lactate dehydrogenase. Patients had a diffuse infiltrate in their lungs visible on X ray. All these symptoms are common in acute respiratory diseases, but the cause of this one was still a mystery. The mortality rate was high, nearly 80 percent, so this was no ordinary disease. Eventually the virus was amplified in the lab by polymerase chain reaction and was found to

Figure 15 The deer mouse, *Peromyscus maniculatus*, is one of the species that carries hantavirus, a disease that, even with good treatment, has a mortality rate of 40 percent. *CDC.*

be a previously undescribed hantavirus. The disease now had a cause—the virus—and a name: hantavirus pulmonary syndrome, or HPS. Now, where was this new virus coming from?

Researchers trapped rodents at the homes of HPS patients in the Four Corners and soon discovered that the deer mouse (*Peromyscus maniculatus*) is the main host of the virus in that part of the country. By mid-January of 1995, 102 people in 21 different states had been diagnosed with the disease, and subsequent examinations of rodent populations revealed four other kinds of hantavirus carried by various species of mice. The disease infects human by inhalation of the virus, which is carried into the air when rodent droppings or urine are disturbed. In North America, the disease does not spread from person to person. There isn't any drug that treats the disease, but with the recognition of the illness at earlier stages and careful fluid replacement in patients with advanced disease, the mortality rate decreased to 40 percent, a significant improvement over the earlier figures, but hardly cause for celebration. There are now trials going on to see if intravenous ribavirin will be helpful in treatment.

Leptospirosis, a bacterial disease carried by rats and other animals, can easily be transmitted to humans through contact with the animal's urine, often present in rat-infested alleys and parks in the inner city and very likely in suburban areas as well. If the bacterium enters your body through an open cut, your eyes, your mouth, or your nose, it can cause a nasty flu-like illness that in about 10 percent of cases becomes systemic. This form is called Weil's disease and can cause fever, jaundice, meningitis, kidney failure, and internal bleeding. On rare occasions, it kills people.

You get the symptoms of leptospirosis about two to four weeks after contact with the bacteria. Since lots of other animals besides rodents can be infected, you can get the disease in many different places. It's an occupational hazard for people who work with animals, and a recreational risk for people who hike, camp, swim, or go boating or whitewater rafting in or near contaminated water. The illness can be successfully treated with doxycycline or penicillin, provided you get it early enough in the course of the disease. In severe cases, intravenous antibiotics are used.

There are about 200 reported cases a year of leptospirosis, which can only be diagnosed with a polymerase chain reaction test of blood or spinal fluid to determine the presence of the bacterium. About half of these cases occur in Hawaii. But the symptoms in most cases are difficult to distinguish from a flu, and few labs perform the complex test for it. So there are probably many more unreported cases of the disease than reported. There was a large confirmed outbreak in 1998 in which more than 100 people participating in triathlon competitions in Wisconsin and Illinois became ill, probably from the polluted water of Lake Springfield.

Ingesting food or water containing rat feces or urine can also give you rat-bite fever. This is a bacterial illness caused by *Streptobacillus moniliformis*, and it can also be contracted by a bite or scratch of an infected rat. The illness usually begins about two to four days after contact with the infected substance. The symptoms can get nasty: relapsing fever, aching bones, and a rash on the extremities, palms, and soles. Headache, meningitis, nausea, and vomiting have also been reported. Although most people get better by themselves, intravenous penicillin for a week, followed by another week of oral penicillin, is a good idea because 13 percent of untreated cases are fatal. Diagnosis is not easy—there's no serologic test, and you can only establish the diagnosis by culturing the bacterium, which is not easy for most labs to do. No one knows if it's effective to give antibiotics prophylactically after an exposure. Fortunately, rat-bite fever is rare.

The species of rat that infests American farms and cities is called *Rattus norvegicus*, or the Norway rat, and this is the one you may see hovering around the garbage cans in the back of the restaurant. It's sometimes called the common brown rat, and it's very ugly—rat-faced indeed, with coarse brown and black fur and a naked tail. They weigh

Figure 16 *Rattus norvegicus*, or the Norway rat. Their babies are pink and hairless, and they have lots of them. When they grow up, they can be a good foot and a half long, if you include the tail. *Colorado State University. Photograph by E. J. Taylor.*

about a pound, and including the six-to-nine-inch tail, they can be a foot and a half long. They have keen senses of hearing, smell, taste, and touch, and they'll eat anything—grain, meat, insects, plant matter. If there's no other food around, they'll eat each other. They need about an ounce of water a day to wash it all down. They can chew through wood, wallboard, plaster, and concrete if they're hungry and there's food available. They can climb, swim, and fit through any opening larger than a half-inch across. They're nocturnal and normally quite shy of humans and other animals (including, usually, each other), but every once in a while they'll bite a sleeping baby, usually on the face where they smell food. A rat lives about 18 months and starts having babies at three months. Six to 12 young are born after about three weeks of gestation, and the average female gives birth about five times a year. So if a female has about 45 babies a year, that means a lot of rats. All they need is enough food to eat and they'll be all over the place.

There is no evidence that the house mouse (*Mus musculus*) you see running across the restaurant floor while you're eating dinner is likely to give you a disease like hantavirus, but rodent droppings and urine are nevertheless something you'll want to keep away from the food you eat. There are some diseases that they transmit, mostly minor, like salmonellosis. And of course some species can be the vector of more serious illness, like Lyme disease. But house mice and humans have lived together, if not always happily, for thousands of years, and continue to

do so today. The two species are quite well adapted to each other. House mice weigh about a half-ounce, and to some eyes they're kind of cute: furry brown things with big ears and small eyes. If a wall is the slightest bit rough, they can grip on and run right up it, and they can jump more than a foot off the floor. A hole about one-quarter inch in diameter is big enough for them to squeeze through. They like cereal grains, but they'll eat almost anything their human companions will eat. They don't live long, but they keep busy: the gestation period is about three weeks, and a female can have as many as 10 litters of five or six young in her one year of life.

There is the occasional, more or less rare disease carried by some species of mouse, and they are the vectors of certain diseases like Lyme, but house mice are in general not a reservoir of serious human illness. They do a certain amount of economic damage—on farms, for example, they can result in the spoilage of large amounts of food meant for humans, and any restaurant that allows them to frolic in the presence of diners is likely to be out of business soon—but, like cockroaches, they and their excreta are probably more disgusting than dangerous.

Flies

Here's a word we hope you've never heard before: myiasis. Myiasis refers to diseases caused by the deposition of insects or insect larvae in living tissue. And this happens in real life, in both animals and man, and not just in horror movies. One Indian boy, it was reported in 2001, actually had a wound on his gum infested with fly larvae, causing a swelling that covered some of his teeth. They treated him with ether, which forced the larvae out, and the wound healed. This is the perhaps the most spectacular way in which houseflies can endanger humans, but there are more mundane ways as well.

Houseflies carry various bacteria and viruses, and they can cause trouble for the humans with whom they inevitably interact. Even killing them can cause trouble: one study showed that electrocuting insect traps—the kind that produce that satisfying buzz every time they kill a fly—may actually spread germs into the air, where you can breathe them right in. Letting them live might be pretty bad, too, depending on where they walk and what that particular fly is carrying around with him. Viruses, bacteria, protozoa—you name it—they're all carried by

houseflies, or, as they have been called, with an eloquence too rare in scientific writing, "refuse and promiscuous-landing synanthropic filth flies."

Filth flies are appropriately named. They love animal feces, and when they visit them they pick up all sorts of stuff. *Cryptosporidium* have been detected both in the digestive tracts and on the external bodies of flies that have feasted on bovine diarrhea—enough oocysts to cause a nice infection in anyone with the bad luck to get them into his mouth. They also can probably transmit *E. coli*. The O157:H7 strain was found in 1997 in some flies from a farm in Japan. One of the most common diseases you can get from flies is *Salmonella* poisoning—having some flies do a soft landing on the egg salad at a picnic is a good way to transmit this bellyache.

Shigellosis can be transmitted by flies, too, thanks to their love of human excrement. Flies become contaminated with *Shigella* when they land on it to feed. Researchers have concluded that fly control, especially in developing countries, is a public health measure that can save many lives, especially those of young children.

Now here's some good news about flies. Since it isn't clear how *Helicobacter pylori*, the organism that causes stomach ulcers, is transmitted, a lot of people suspected flies might be involved. But in this case, if in this case alone, they're not guilty. Some researchers ran an experiment by exposing flies to human feces infected with *H. pylori*, but the flies just didn't pick it up. How this bacterium is transmitted is still an unsolved mystery, but flies have nothing to do with it.

Dirty Money

Rats, mice, flies, cockroaches—these are just things you can run into at almost any restaurant at one time or another. Every time you go out, you're exposing yourself to something. In fact, those dollar bills in your wallet, the ones you plan to spend for the taxi ride downtown before you even do anything, are covered with germs. According to ABC News, some researchers at Wright Patterson Medical Center in Dayton, Ohio, decided to take a look at some dollar bills changing hands at a grocery store and at a high school concession stand. They asked people to exchange their dirty wrinkled bill for a nice new one, and found that 87 percent of the bills they collected were contaminated with bac-

teria that could conceivably cause infection in someone, either healthy or immunocompromised. Of 68 dollar bills examined, only 4 came up relatively clean. The meaning of this isn't entirely clear, of course—whether you're actually spreading contagious disease every time you buy a newspaper is another question entirely. Still, there it is: very dirty money.

Gym Risks

We all know it's healthy to go to the gym regularly, but did you ever wonder what's in the sweat of the guy who used the exercise machine just before you? Sure, he wipes off the machine with that towel, but that's not exactly sterilizing, is it? Are you supposed to touch that stuff, and what happens if you do? Does sweat have germs in it?

No, sweat does not have germs in it. When it comes out of your skin, in fact, it's sterile and completely odorless. But soon after it emerges, stuff starts to grow in it, and it's these microscopic organisms that make sweaty people stink. Bacteria and fungi produce enzymes that result in the decomposition of organic matter on the skin—the lipids of the sebum (fatty secretions from certain skin glands) and the protein components of the skin itself. Sweat is one place where these processes can take place, and thus sweating leads to what doctors call bromidrosis, and what the rest of us call b.o. Certain diseases such as diabetes, gout, and scurvy can make sweat smell worse, and so can the ingestion of some foods or medicines.

A wet, warm armpit or a cozy place between the toes is a good place for bacteria, yeasts, and fungi to grow—those places smell bad for good bacterial reasons. Some researchers discovered that the quality of the odor depends on the quantity of bacteria—"pungent" body odor is caused by greater numbers of bacteria than "acid" body odor. The worse you smell, in other words, the more bacteria you're growing. In any case, everyone is growing plenty of them—half a million to a million organisms per square centimeter of underarm. And it's not just any bacteria. Only certain gram-positive bacteria called aerobic diphtheroids cause that distinctive odor. This, of course, leaves quite a few different species that can stink the place up. Some people think men smell worse than women, and they're probably right—in this same study, men had larger quantities of diphtheroids than women, and propionobacteria were

recovered from 70 percent of males, but only 47 percent of females. Other species of bacteria were found in roughly the same quantity in either sex. The difference in smell was borne out by observation: among 20 males, the ratio of "pungent" to "acidy" was 17 to 3, while only 6 of the 13 women were "pungent." In case you were wondering, and you almost certainly were not, the scientists didn't find any difference between the amount of bacteria in the right and left armpits of either right- or left-handed volunteers.

Can touching a sweaty piece of exercise equipment transmit disease? Theoretically, yes. If you put your hand in some sweat, and there were bacteria in it, and you then put the same hand in or near your mouth or nose, you could conceivably catch something. As you know, you can catch something just by shaking hands with another person. But that sterile sweat as it falls off a human body is unlikely to carry much in the way of bacteria, since bacteria take time to multiply to a point where they can transmit illness. People should wipe off exercise equipment when they're finished with a workout. They might prevent disease transmission, but more important, they will certainly prevent revulsion in the next person who uses the machine. Whatever tiny health risks there may be in touching another person's sweat on an exercise machine at the gym, they are far outweighed by the health benefits of vigorously using that machine, sweaty or not.

Nailed at the Salon

Ever get your nails done at a nail salon? A lot of people do: nail care is a $6 billion a year industry in the United States, and there are tens of thousands of nail salons all across the country—7,500 in California alone. One day in 2001 a dermatologist called the health department and reported that he had five patients with infections of the feet and legs that weren't responding to antibiotics. All of them had been customers of the same nail salon in Watsonville, California. The health department was a bit concerned, so officials paid a visit to the salon in question. Before long, they'd found the problem.

Seems the footbath in which people were soaking their feet for 10 or 15 minutes per session was contaminated with *Mycobacterium fortuitum*, a bacterium closely related to the one that causes tuberculosis. One hundred ten customers, all but one of them women, were infected

in this way after soaking their feet, and one of the women had 37 boils on her feet and legs. Some of the women got so badly scarred they needed skin grafts, and medical costs ran as high as $10,000. *M. fortuitum* is ubiquitous, and it sometimes causes human infections, but this was the first time it had ever caused such a widespread outbreak of disease. The salon owner said he hadn't cleaned or disinfected his footbaths since he installed them a year before, and the result was a buildup near an inlet screen of hair, grease, and clipped nails that formed a perfect culture medium for the bacteria. The *New York Times* quoted Dr. Kevin L. Winthrop, who did the investigation: "There was enough hair to make a toupee." Doxycycline and ciproflaxin are usually effective in combating *Mycobacterium* infections, but some of these women had been on antibiotics for months without having the problem clear up. California health officials immediately began preparing to issue regulations about cleaning and disinfection of footbaths.

Hospitals

You go to a hospital to get well, but you have to remember that a hospital is one public place that's bound to be filled with germs—after all, it's people with germs who wind up there, all kept close together with a staff of caretakers who are almost inevitably going to transmit germs from one of their patients to another. And we're not even mentioning the visitors who come in and out dragging all sorts of microscopic organisms along with them. This kind of disease—the kind you pick up in a hospital—is so common that there's a name for it: nosocomial infection. Of course, nosocomial infection can be avoided, but it's not easy, and it requires the strict cooperation of everyone from the janitor who mops floors to the surgeon who transplants hearts.

Every doorknob in a hospital is a source of possible transmission of disease, since people touch them and people's hands are covered with germs. Modern stainless steel doorknobs and doorplates look nice and shiny, but they may be less hygienic than the old-fashioned (and slightly less expensive) brass hardware. Stainless steel, which is 88 percent iron and 12 percent chromium, is a much better place for germs to live than brass, which is about 67 percent copper and 33 percent zinc. Brass, unlike stainless steel, is actually germicidal, even when it's tarnished. Of course, doorknobs are only one of the places where bacteria can thrive in a hospital.

The recommended procedures for sanitizing patient care areas of hospitals are somewhat involved. Different techniques reflecting different levels of sanitation are used for cleaning medical equipment, housekeeping surfaces, operating rooms, and so on. Floors of hospitals, for example, don't need any kind of extraordinary cleaning or disinfecting. Several studies have shown that disinfecting hospital floors has no advantage over simply mopping them with detergent and water. The best approach is to use a technique that doesn't stir up dust, and to change cleaning solutions and wash cleaning implements often. Letting mops and buckets dry between uses is also helpful, and cleaning solutions should be discarded after use, not saved for reuse. This doesn't make the floor germ-free, since any floor that has been cleaned immediately becomes contaminated by shoes, the wheels of equipment, body substances, and airborne microorganisms.

On the other hand, areas where immunocompromised patients or newborns are housed require elaborate precautions. Here, surfaces have to be wet-dusted with disinfectants, always making sure to avoid stirring up dust near patients or allowing patients to come in contact with cleaning fluids. Cleaning equipment that produces mists or aerosols cannot be used in these areas. Vacuuming is a problem—it can spread microorganisms through the air, and has to be done carefully using the right kinds of filters. The filters have to be cleaned and changed regularly, depending on the kind of vacuum cleaners used. Infant beds have to be cleaned and disinfected, and infants frequently moved to clean bassinets or incubators. Any equipment has to be washed thoroughly with water and allowed to dry before being used.

Although there is no evidence that anyone has ever been infected with hepatitis B, hepatitis C, or HIV by contact with a contaminated environmental surface, blood and other bodily fluid spills have to be cleaned and the area promptly disinfected. HIV is inactivated very rapidly after being exposed to commonly used germicidal agents, including ammonia solutions, and even a 1:100 dilution of ordinary household bleach. Embalming fluids like formaldehyde also completely inactivate HIV and HBV. When large volumes of infected blood or body fluids are spilled, cleanup must be more extensive and decontamination more thorough. In such cases, all organic material must be removed by a person wearing gloves, and a 1:10 solution of bleach is required.

Carpeting in hospitals has advantages (it reduces noise and makes the place a bit more homey), but it's a good place for bacteria and fungi

to grow, and cleaning it can present problems. Its use should be avoided in areas where spills are likely to occur. Vacuuming, both dry and wet, can disperse bacteria into the air, and carpeting is more or less constantly infected with bacteria, even right after cleaning. Some hospitals use carpet tiling, discarding and replacing tiles that have become infected by a spill.

Even flowers can bring unwanted organisms into a hospital room. They can be infested with bacteria when they're first brought in, and the water in which cut flowers are kept can become a breeding ground for microorganisms of various kinds, usually gram-negative bacteria like *Pseudomonas*. Spores of fungi such as aspergilla and fusarium are also commonly found in both fresh-cut and dried flowers. But studies have shown that flowers in a hospital are no more infested with bacteria than flowers in people's houses, in restaurants, or anywhere else. Flowers are germ-laden wherever they are, but they don't really pose a problem for patients whose immune systems are working properly.

Even though they are rarely implicated in nosocomial disease outbreaks, cockroaches, flies, and mice are no more welcome in a hospital than anywhere else, and they can be a threat to susceptible patients. *Shigella* carried by houseflies, for example, may be one cause of diarrhea in pediatric wards, and diarrheal infections are potentially life-threatening in infants.

Perhaps the biggest threats for infection control in hospitals lies in methicillin-resistant strains of *Staphylococcus aureus* and vancomycin-resistant enterococci. More than half of the staph infections in intensive care units are methicillin resistant, and about a quarter of enterococcal infections are resistant to vancomycin. Although you can find these bacteria on environmental surfaces, that's not how they are transmitted. They almost invariably go from person to person.

Viruses are a problem everywhere in a hospital, but particularly so in pediatric and neonatal wards. Respiratory viruses including rhinoviruses, respiratory syncytial virus, adenoviruses, influenza viruses, and parainfluenza viruses present significant problems and are usually transmitted by aerosolized small particles or by hand contamination with respiratory secretions. While most viruses can only persist for a few hours on environmental surfaces, that's often long enough to transmit them to new victims. Enteric viruses (e.g., rotaviruses, adenoviruses, and caliciviruses) are usually passed along by the fecal-oral

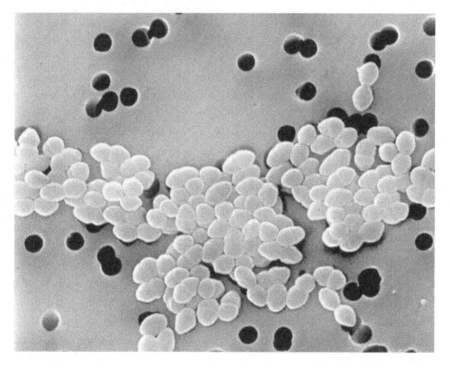

Figure 17 A scanning electron micrograph of vancomycin-resistant enterococci. Small changes in surface proteins can make bacteria resistant to the most powerful antibiotics we have. Vancomycin is usually the antibiotic of last resort, but it just doesn't work against these organisms. *CDC. Photograph by Janice Carr.*

route. These affect both patients and staff and are especially troublesome where patients have diarrhea. Careful cleaning of environmental surfaces and frequent hand washing are the only sure ways to prevent spread.

So what's a poor hospital patient to do? Since washing hands often is one of the best ways to prevent the transmission of germs from one patient to another, some have recommended the direct approach: ask all staff members who approach your bed whether they've washed their hands, and don't let them touch you unless they have. While some will be comfortable with this, to others it may sound like a good way to infuriate people, especially those who actually have washed their hands, and not a practical method of assuring conscientious care. Of course, if you are under anesthesia or otherwise unable to pose such questions, you just have to rely on the staff assigned to your care. Other than choosing a hospital with good infection control practices—not always easy

for the average person to determine, although every hospital has written guidelines that you can read—you are really at the mercy of the people who are taking care of you.

Fellow Travelers

A lot of people worry about "picking up something" while they're using public transportation. In New York City, the most feared location for mysophobes is undoubtedly the subway. Everyone squashed together, sneezing, coughing, wiping their noses, and then grabbing those poles to keep from falling over every time the train stops, starts, or lurches. The *New York Times* was so convinced people worried about this that they actually went out and hired an infectious disease specialist, Dr. Aaron Glatt, chief of the department of infectious diseases at St. Vincent Catholic Medical Centers Brooklyn and Queens, to go around the subway with a bunch of Q-Tips taking swabs to see what he could find. This was not, the doctor emphasized, a scientific test, and any results would have to be viewed in light of the casually gathered small sample, which would say little about the subway system as a whole. But he did go into the F line in Queens and take about 15 swabs from various places—handrails, poles inside the cars, a pay phone receiver, the handrests on a wooden bench. Then he cultured the material on agar plates. What he found was reassuring: nothing except some ordinary *Staphylococcus*, the kind of staph that just about everyone has on his skin and that doesn't harm anyone. The subways may not be all that clean—they certainly don't look too clean—but they're probably not a more significant source of infectious disease than any other public place where people touch things a lot.

Public Bathrooms

There are lots of rumors and frightening stories that go around, but some of them just aren't true. You can't get syphilis from a toilet seat, no matter what you may have heard, and HIV is not transmitted by contact with environmental surfaces in public bathrooms. You can't get herpes from contact with toilet seats, either—you only get it by close contact with someone who is already infected.

So what can you really catch? Plenty. A bathroom, public or private, is a good place to make contact with human feces, and as you already know, there are dozens of unpleasant diseases that can be transmitted by such contact. While most germs don't survive too well on environmental surfaces, some survive long enough to be transmitted. These, of course, are the ones to worry about.

Cryptosporidium can be transmitted by contact with bathroom fixtures or changing tables in public bathrooms, and you can wind up with a bad case of diarrhea from such contact. *Giardia, Salmonella, Clostridium,* and *Shigella* can be transmitted in the same way. There is at least one study of *Salmonella* which showed that the germ can persist for weeks after someone with diarrhea uses the toilet, and flushing the toilet can cause contamination of the toilet seat and toilet seat lid. While the germs didn't survive on surfaces that got dry (sinks, water faucets, etc.), they did stick around under the rim of the toilet, where someone could conceivably come in contact with them. Bathrooms used by diaper-clad kids with diarrhea are a particularly germy place. Norwalk-like viruses can also be spread in public bathrooms. One study found that shared bathrooms on college campuses were a source of the spread of this unpleasantness. But you have to do more than just touch the stuff: you have to get it in your mouth or up your nose.

Are sparkly-clean bathrooms safer than ones that are ill maintained? They're certainly a lot more pleasant, but that doesn't mean they're less likely to make people sick. The amount of contamination in the bathroom is much less important in disease transmission than the germs on your hands when you leave the bathroom. A 1983 study in Houston found that in a day care center there was little difference in the amount of contamination found in bathrooms during a diarrhea outbreak and at any other time—the big difference was that at times of diarrhea outbreak, more hands and more toys were contaminated. In other words, it was the lack of hand washing, not the level of contamination of the bathroom, that was causing the outbreak of disease. Whatever the level of cleanliness of the bathroom, and whether it's in the Waldorf Astoria with flowers on the sink or in back of the gas station with a broken toilet seat, you're much more likely to get into trouble if you don't wash your hands after visiting it.

The solution is simple: wash your hands thoroughly before you touch your nose or mouth. You don't have to scrub up like a surgeon preparing for an operation—ordinary hand washing with plain soap

and water is enough. This precaution will reduce the risk to nearly zero. Perhaps the best advice to follow is this: if the public toilet isn't equipped with soap, hot water, and paper towels, don't use it. And if it is so equipped, take full advantage of all of the facilities. This advice may be difficult to follow in certain instances: when using an outhouse during a camping trip, for example, or when traveling in exotic places. But even then, finding a place to wash your hands is a sensible thing to do before you do anything else.

Cruising

On July 15, 1994, the New Jersey State Department of Health reported to the CDC that six people who had traveled to Bermuda on the cruise ship *Horizon* had come down with pneumonia. The CDC quickly discovered that 14 people on the ship had been infected with Legionnaire's disease. It was pretty clear that the disease had come from the same place—somewhere on the ship. They began an investigation of the water supply system aboard the liner and quite quickly found the culprit: *Legionella* species were found in the whirlpool bath, and they were the same subtype as the type found in the 14 ill passengers. *Horizon* immediately hyperchlorinated its water supply, removed the whirlpool filters, and discontinued use of the whirlpool baths. The ship resumed its regular schedule on July 30. This was the first documented case of Legionnaire's disease aboard a cruise ship.

From August 29 to September 1, 1994, 1,589 passengers took a round-trip voyage on the *Viking Serenade*, owned by Royal Caribbean Cruises, Ltd., from San Pedro, California, to Ensenada, Mexico. Five hundred eighty-six passengers and 594 crew members wish they hadn't. They all came down with diarrhea and vomiting, and it had nothing to do with seasickness. The cause was *Shigella flexneri*, which was isolated from stool samples of ill passengers. One man, 78 years old, died in Mexico with diarrhea. The next two scheduled cruises of the ship were canceled while the CDC began an investigation of the source of the outbreak.

It was earlier instances of problems like this that prompted the CDC in the early 1970s to begin a program of inspections of cruise ships called the Vessel Sanitation Program, or VSP. In the 1970s and 1980s, there were an average of 12 to 15 outbreaks of diarrheal illness

on board cruise ships every year. The VSP quickly began to have its effect, and by 1999 there were only three outbreaks.

The program works like this. First, it's voluntary. About 140 ships are signed on to the program. Each ship undergoes an unannounced inspection twice a year while it is in a U.S. port. Each ship starts out with 100 points, and points are deducted for various kinds of violations. If a ship gets an 86 or lower, it will be reinspected within a month to six weeks. Depending on the size and complexity of the ship, the inspections may involve two to four inspectors and last five to eight hours. The inspection covers the ship's water supply; spas and pool; food preparation, storage, and service areas; personal hygiene of employees; and general cleanliness to ensure the absence of insects and rodents. It also examines the ship's public health training programs to determine their effectiveness.

You can look up scores at www2.cdc.gov/nceh/vsp/vspmain.asp. A number of ships get perfect 100s, most score in the 90s, and then there a few that fail, as did the *Celebration* in November 2001. This ship, owned by Carnival Cruise Lines, Inc., lost 15 points for a large variety of violations in its water system, food preparation procedures and equipment, dishwashing equipment, cleanliness of food storage areas, and so on. While the majority of ships do pretty well, sometimes some very fancy ships get some very bad scores. The *Queen Elizabeth 2*, for example, failed inspections with scores of 80 in April 1997 and 79 in January 2000. It hasn't had a perfect score in any inspection since 1992, as far back as the listings go. In its most recent inspection, on July 31, 2001, it scored an anemic 89. The ship lost five points for potentially hazardous preparation, display, service, and transportation of food; two points for errors in food storage; two points for keeping food at the wrong temperature; one point for poor maintenance of nonfood contact surfaces; and one point for dirty bulkheads. The ship issued a statement to the VSP as required, listing corrections of 55 specific points that the VSP wanted addressed.

What can you do to protect yourself on a cruise? First, remember that outbreaks of disease on cruise ships are extremely rare—even on those that get low scores or fail inspections. You could check the records of your ship—the web site lists inspections over the last nine or 10 years, and this might give you some idea of the ship's hygiene maintenance over time. Beyond that, probably the most important thing you can do to prevent outbreaks of diarrheal or other infectious diseases on

Figure 18 The *Queen Elizabeth II*, according to Cunard, "the most prestigious ship in the world." You can't make out in this photograph the dirty bulkheads, the food kept at the wrong temperature, or the poor maintenance of nonfood contact surfaces. The potentially hazardous preparation, display, services, and transportation of food is also impossible to detect. The *QEII* hasn't had a perfect score from the Vessel Sanitation Program in more than a decade.

a ship (as well as anywhere else) is to wash your hands frequently. It is probably safe to say that on most ships your risk of getting seasick is considerably higher than your risk of any other kind of stomach upset.

Flying

What happens when you spend six or eight hours in an airplane with 300 other people, breathing recirculated air? People sometimes believe they've been made sick by flying in an airplane, but in fact it's difficult to prove that anyone is any more likely to contract an infectious disease spending three hours in an airplane than spending the same three hours in any indoor space with other people. Passengers, flight attendants, and pilots often complain about the effects the air in an airplane has on them, but those nonspecific symptoms are very rarely caused by germs. In fact, the Department of Transportation and Consolidated Safety Services did several studies that found that the only organisms detected on 35 flights of eight different U.S. airlines were normal skin flora, with no respiratory pathogens at all. They even found that bacterial counts were somewhat lower than they were in public buildings.

Large commercial airliners are required by government regulation to keep pressure at the equivalent of an outside air pressure of 8,000 feet. This is a comfortable air pressure for almost everyone to breathe

normally. Some people believe that when your're in an airplane you're "breathing the same air over and over," but this isn't so. In older airliners, such as DC-9 and the 727, air is drawn from the outside of the plane, compressed to cabin pressure, cooled, circulated, and exhausted. In new airplanes, half of the cabin air is recirculated. Cold air from the outside enters air compressors controlled by the plane's jet engines. Compressing air heats it up, so it must be cooled before it can be used. The air is cooled first by heat exchangers and then, after flowing through ducts in the wings, cooled further by air conditioning units located underneath the cabin floor. The cooled air is mixed half and half with cabin air and then pumped through filters to overhead outlets. Meanwhile, air is drawn off by grilles located in the cabin's walls or ceiling. Half of this air goes into the air mixer and the other half is exhausted outside the airplane, through a valve that also controls air pressure inside the plane. Mixing outside air and cabin air is more economical for the airline—it saves fuel. Some complain that changing only half the air and then mixing it with cabin air is less healthful than changing all of it, but there is little scientific evidence to support this view.

The filters through which the air travels, high-efficiency particulate air (or HEPA) filters, are not much different from those used in hospitals, and they can trap particles as small as bacteria and even viruses. Odors dissipate very quickly because the air is exhausted and replaced 20 to 30 times an hour, depending on the airplane, with a combination of outside air and filtered air. This is somewhat more often than air is recirculated in large office buildings, but of course the population density is much higher and the volume of air available per person is much smaller in the confined space of an airplane cabin. Bringing in a constant supply of outside air replenishes oxygen, while the exhausted air dissipates carbon dioxide outside the plane. The recirculated portion of the air is constantly diluted with outside air. On the whole, the system works very well and keeps almost everyone comfortable and healthy.

Still, there have been some unpleasant incidents of infectious disease transmission aboard airplanes, even though an airplane's air circulation system has never been implicated in any of them. Several instances of exposure to tuberculosis on airplanes are illustrative. Between January 1993 and February 1995, the CDC investigated six instances in which passengers or crew flying on commercial airliners traveled while infected with TB. All six were symptomatic with the pul-

monary form of the illness and were highly infectious at the time of the flights. Two of them had a strain that was resistant to both rifampin and isoniazid—the most serious kind of drug-resistant TB. Two of the patients were aware of their condition, and in fact were on their way to the United States for treatment.

In one case, a flight attendant had had a positive skin test for TB but continued to fly anyway. She developed an increasingly severe cough and was diagnosed with pulmonary TB in November 1992. The investigation revealed that the flight attendant had transmitted TB to a number of members of the flight crew, but probably not to passengers. The longer the flight crew flew with her, the more likely they were to be infected. The type of airplane they flew in didn't matter, but if they spent more than 12 hours with her as a crew member, they were much more likely to have positive TB tests. In another case, a passenger with TB traveled in the first-class section on a flight from London to Minneapolis that lasted nine hours, but he didn't transmit anything to anyone. In a third case, a TB patient flew from Mexico to San Francisco, and when the San Francisco health department investigated, they found that 10 people had positive TB tests of the 22 they were able to contact and test. But even in this case, it wasn't clear that they had contracted the disease during the flight: nine of them were born outside the United States, and the tenth was a U.S. citizen who may have been infected outside the country. The health department found was no conclusive evidence that the illness had been transmitted during the flight. In only one of the six cases was there clear evidence that the disease had been transmitted during the flight. This was a flight in April 1994 after which six people appeared to become infected. Four of these were seated near a foreign-born passenger with pulmonary TB on a nine-hour flight from Chicago to Honolulu.

In all cases where transmission of tuberculosis occurred in an airplane, the infection was passed person to person to nearby passengers during long flights. Essentially, this means that your chances of contracting TB on an airplane are no different from what they are in any confined space—you have to be exposed directly to an infected person for a significant length of time. *M. tuberculosis*, even though it is an airborne bacterium, is not carried to other people through the plane's air circulation system.

The bacterium *N. meningitidis* is the leading cause of bacterial meningitis in the United States, and it can be deadly. Of 2,600 people

who get it every year, 13 percent die. The CDC believes that meningo-coccal disease might theoretically be transmitted on airplanes, even though there have been no confirmed instances of this happening. Still, anyone who sits near someone who is infected is at greater risk for infection. Antibiotic prophylaxis is recommended only for those who have direct contact with the infected person's respiratory secretions or who sat next to the person for eight hours or more. Prophylaxis is also recommended for household members traveling with the person and other close contacts, such as roommates or members of the same sports team, but not for any other passengers. Again, the bacteria are not transmitted through the air circulation system of the airplane.

An outbreak of flu on an airplane in 1979 suggests that an airplane's ventilation system, far from passing on infections, actually prevents them from being passed on by filtering the air. The flight was delayed on the ground because of an engine failure during a takeoff attempt. While repairs were being made the engines and the ventilation system were inoperative. Most passengers stayed on the plane for three hours. Within four days of the flight, 72 percent of the passengers had come down with cough, fever, fatigue, headache, sore throat, and myalgia. In other words, the flu—in this case identified as A/Texas/1/77(H3N2).

Some countries require planes to be sprayed with insecticide while the passengers are aboard. These countries are located in Latin America and the Caribbean as well as in Australia and the South Pacific region. There is good reason for this, because killing insects, particularly mosquitoes, before they are imported to another country is an important health measure. Some people may be adversely affected by the spraying—those with allergies, for example. There is a high likelihood that mosquitoes brought in by airplane have introduced malaria into regions where the disease was previously unknown. This happened in Brazil in 1996. Three people were diagnosed with malaria after taking a plane flight from Lebanon to São Paulo on August 16. The trip involved a half-hour stopover in Abidjan, Ivory Coast, Africa, which is apparently where some nonhuman passengers came on board. During an epidemiological examination of the airplane, four mosquitoes were found—one in the luggage compartment and three in the first-class cabin. All three patients had been traveling first class. The species was *Anopheles gambiae*, predominant in Africa. The three recovered, but not before two of them had become seriously ill.

Cholera is a very rare disease in the United States, but the few cases that occur are almost all attributable to travelers coming into the coun-

try, either by airplane or ship, from countries where the disease is endemic. The cholera now being brought into the country, even though it is in very small numbers, is often resistant to antibiotics, but antibiotics are, fortunately, not essential in treating the illness. The only cholera vaccine available in the United States has been discontinued because it was not highly effective. There are, however, two cholera vaccines available in other countries that seem a little better, but they are not recommended for travelers. Cholera is transmitted by the fecal-oral route, not through the air. Most infected people have no or very mild symptoms. Travelers to Africa, Asia, and Latin America should be aware of the risk and avoid consuming suspect water or food.

Pins and Needles

A tattoo parlor doesn't seem like a particularly salubrious place, and having someone stick a needle through your skin seems obviously to carry some risk, but the risk of getting any blood-borne disease by getting a tattoo are extremely low, at least in the United States. Although it is of course theoretically possible to transmit such diseases as hepatitis or HIV with tattoo needles, the CDC has never heard of such a case. Body piercing, which often involves tissue that takes a long time to heal, also involves a theoretical risk if unhealed sores come into contact with infected blood, although there are no documented cases of this kind of transmission, either. There is one documented case of HIV transmission from acupuncture, but there are no data at all to indicate that people who get tattoos or body piercings are at increased risk for hepatitis C or HIV. There have been a few cases of apparent transmission of hepatitis B in tattoo parlors, but no study shows getting tattoos to be a risk factor for the disease.

Of course, in most jurisdictions tattoo parlors and body piercing establishments are pretty much unregulated, so you're taking your chances when you enter one of them. The people who do tattoos don't necessarily have any training in infection control, and the dyes they use are not approved by the FDA. If you're determined to get a tattoo, New York City may be one good place to get one. The New York City health department licenses tattoo artists (not tattoo parlors), who have to take a course in infection control and pass an exam. The law took effect in 1997, and 500 licenses have been issued since then. The license is good

for two years and then has to be renewed. The city does examine tattoo parlors, but only if there's a complaint. The Department of Health reminds everyone that it makes judgments only on health matters—complaints about the quality of the artwork are not within their area of expertise, and the licensing exam does not test competence in artistic expression.

Thanks to childhood vaccination, tetanus is a rare disease in the United States—there are about 40 cases a year. Between 1995 and 1997, the CDC attributed one of 123 cases to a self-performed tattoo and one other to a self-performed navel piercing. There were no cases of tetanus attributed to tattoos or body piercings done in professional establishments.

There are some real dangers out there, and taking certain precautions certainly makes sense. But at least in my view, you're being irrational if you let fear of contagion interfere with the normal pursuit of everyday social interactions, or if your idea of taking precautions is to stay inside your house all day wearing a surgical mask. I am nevertheless aware that not everyone who has read this chapter will share my opinion.

12

The Antiseptic Supermarket

PRODUCTS THAT DO SOMETHING,
PRODUCTS THAT DO NOTHING, AND
PRODUCTS THAT ACTUALLY DO HARM

The company that makes Lysol announced in a television commercial that they are now selling a kind of paper towel soaked in liquid disinfectant—"kills 99.9 percent of germs"—the same percentage that Tide with Bleach (see chapter 4) claims to wipe out.* There are all sorts of products, hundreds of them, that are advertised as antibacterial, including kitchen cleansers, bathroom cleansers, floor and window cleaners, hand cleaners, antibacterial soaps, hoses, meat slicers, paints, even antiseptic toys. There are antibacterial plastic food storage containers and knives, antibacterial mattresses, toilet seats, towels, pillows, sheets, slippers, earplugs, plastic water filters, and chopsticks. You can drop all your antibacterial purchases into an antibacterial shopping cart—a company called Rehrig International manufactures a model whose plastic is embedded with an antibacterial substance. Antibacterial

*The phrase "kills 99.9% of germs" sounds like a classic case of advertising baloney, but it isn't, at least not completely. It actually refers to a scientific measurement in which 99.9 percent of a standard concentration of germs on an agar plate or other culture medium are killed by the application of a certain concentration of a given antimicrobial substance. That's why the FTC lets them get away with saying so. It does not, however, mean that everything you touch with these paper towels will immediately have 0.1 percent of the germs it had before you touched it, although the advertiser will probably not object if you believe this.

claims are now made for more than 700 different products on the market—up from about a dozen in the mid-1990s.

But whatever the advertising for these products may say or imply, there is no scientific evidence anywhere that demonstrates that there are any health benefits whatsoever to using antibacterial cleansers in a healthy household. On the other hand, there is considerable evidence that the products change the microbial flora of a household, allowing resistant strains of harmful bacteria to flourish, and there are increasing reports of an association between too much hygiene and increased allergies. It may be that excessive hygiene interferes with the normal development of the immune system because it eliminates the stimulation caused by ordinary microflora.

To develop normally, the immune system has to create a balance between the T-helper 1 (TH-1) cells, which provide cellular immunity, and the TH-2 cells, which promote antibody production. People with allergies and eczema have an imbalance between the activities of these two kinds of cells as compared with control groups. In people with allergies, antibody production seems to predominate over cell-mediated responses. It may be that an "antigenically rich environment" (that's a fancy way of saying "a dirty house") may be healthier than one persistently sanitized with antibacterial cleansers.

Glenn Cueman, the president of Microban Products Company, a firm that impregnates many products with triclosan (an antibacterial agent), said in 1998 that "there is no real-world evidence of a link between resistant bacteria and triclosan." It's hard to know what Mr. Cueman means by the "real world," but he was probably wrong when he said this, and he is almost certainly wrong today. In fact, there are numerous studies that tie the use of triclosan to the development of resistant strains of *S. aureus*, *E. coli*, and other germs, specifically because it causes a mutation in a certain gene of these bacteria. This is too bad because triclosan is a useful chemical that will become less and less useful as more resistant strains develop. What there really is "no real-world evidence of" is any health benefit from embedding triclosan in consumer products.

The development of resistant strains of microbes is a well-known phenomenon in hospitals, and the strains developed vary from one hospital to another. Even hospitals in the same city can have very different patterns of antibiotic resistance, depending on how antibiotics are used. There is at least some evidence that this can happen in homes where antibacterial products are extensively used, and in any case the theo-

retical risk is obvious. Some doctors believe that antibacterial soaps, for example, while they may be useful for hand washing, should not be used for bathing, where they will destroy useful bacteria along with the harmful, possibly leading to susceptibility to previously harmless microbes. Nevertheless, there are plenty of antibacterial bath soaps available for anyone who wants to use them.

Antibacterial soaps and cleansers may cause other problems around the house as well, specifically in septic systems. A properly installed and well-operated septic system is a very effective way to recycle waste, but it depends on bacteria for its effectiveness. The bacteria break down waste products, making them harmless or even helpful. Anaerobic bacteria in the septic system help decompose waste matter, and aerobic bacteria in the soil render pathogens harmless. In other words, these are bacteria you don't want to kill. Yet antibacterial household products do just that—kill helpful bacteria along with the harmful. The University of Minnesota Extension Service recommends that "every flush" toilet cleaners not be used with septic tanks, and that only the smallest possible amounts of antibacterial soap, drain cleaners, toilet bowl cleaners, laundry products, tile cleansers, and tub cleaners be flushed into the system—and this includes products labeled "safe for septic systems." The use of one product one time may be "safe," but the cumulative effect of using many of them often can be disastrous. The university recommends further that chlorinated water from swimming pools never be introduced into a septic system, and that unwanted medications never be flushed down the toilet. All this is to preserve the bacteria that do the job of recycling waste products.

Label Lexicon

Why do they print directions on the back of your liquid antibacterial hand soap bottle? Do you really need "directions" on how to use soap? Antibacterial products used to clean bathrooms or kitchen sinks are not drugs, and therefore their labeling is regulated by the Federal Trade Commission. Same for ordinary (i.e., nonantibacterial) soap. But if a product intended for use on or in humans claims that it is antibacterial or makes any other medical claim (such as that it will cure dandruff, grow hair, or prevent acne), then the Food and Drug Administration classifies it as a drug, and it falls under their purview. (The Environmental Protection Agency has a role in this regulation as well, as we will see later.) If a manufacturer

wants to sell a drug, even an over-the-counter drug, he must follow many regulations that involve a great deal of reporting, registering, renewing registrations, submission of forms, complying with standards, obeying quality control and labeling requirements, undergoing periodic inspections, and assuring that the ingredients are both safe and effective for the purpose they are supposed to accomplish. It is not a procedure that can be taken lightly, so all of the companies that sell antibacterial soap or hand cleaners must spend considerable time, effort, and money before they can do so. They obviously feel that it is worth the trouble.

The labeling requirements are quite demanding. Every label has to say what antiseptic is in it. It must include indications for when it is to be used, and directions for proper use. It must include warnings like "for external use only," or "do not use in eyes," or "discontinue if redness or irritation develop." Manufacturers are not allowed to use "antibacterial" as an indication, although they are allowed to make general statements like "Contains antibacterial ingredients." Approval by the FDA also requires that the product be tested and found effective, and must include animal and *in vitro* testing. The testing also has to determine how quickly the drug achieves its antimicrobial effect, and all this must be reported to the FDA. When an antibacterial hand soap boasts in its advertising that it has been "scientifically tested and found effective" or "approved by the FDA," the manufacturer is making a virtue of necessity. Colgate Total toothpaste, for example, claims in its advertising that "Colgate Total is the ONLY toothpaste approved by the FDA and accepted by the American Dental Association to help prevent plaque and gingivitis." That's true, and Colgate can say so, because in 1997 the company received approval from the FDA to include the antibiotic triclosan in the toothpaste. Other makers of toothpaste are free to do the same thing. Colgate Total toothpaste, as far as the FDA is concerned, is an over-the-counter drug. On the labels of antibacterial products, you will see an 800 number that you can call "for questions or comments." The manufacturer may have set up this number out of the goodness of his heart, but in any case the FDA requires the company to do so and requires it to have a live person answering it.

Truth in Advertising

Although the FTC doesn't regulate the production or labeling of over-the-counter drugs, it does have a say in the way they are advertised. And

they can be just as fussy as the FDA when it comes to making claims about antibacterial products. A complaint the FTC filed against Unilever in 1999 is illustrative.

Unilever manufactures Vaseline Brand Intensive Care Antibacterial Hand Lotion. They advertised it as "antibacterial hand lotion" that "stops germs longer than washing alone" and "stops germs for hours." The FTC cited several different print and TV ads that made these claims and maintained that Unilever had expressly or by implication said that:

A. Vaseline Brand Intensive Care Antibacterial Hand Lotion stops germs on hands longer than washing alone.
B. Vaseline Brand Intensive Care Antibacterial Hand Lotion provides continuous protection from germs for hours.
C. Vaseline Brand Intensive Care Antibacterial Hand Lotion is effective against disease-causing germs, such as cold and flu viruses.

The FTC begged to differ on all three claims. There is no scientific evidence that the lotion stops germs on hands longer than washing alone. Triclosan, the active ingredient in the hand lotion, can reduce the number of germs on a user's hands, but the degree and duration of its effectiveness have not been established scientifically. And finally, triclosan has no effect whatsoever on viruses, which cause colds and flu, even though Unilever was implying that it does. The FTC viewed all this as false advertising and unfair or deceptive acts or practices in violation of various sections of the Federal Trade Commission Act.

After some discussion among Unilever lawyers and the FTC, Unilever, however reluctantly, began to see things the FTC way. Without admitting that they had ever done anything wrong in the past, they agreed in the future to stop doing those things they never admitted having done. They agreed to stop saying that:

A. Such product is as effective as, or more effective than, washing alone in protecting users against germs;
B. such product has a continuous effect against germs;
C. such product has any effect on any specific germ;
D. such product treats, cures, alleviates the symptoms of, prevents, or reduces the risk of developing colds, allergies, influenza, food-borne illnesses, or any other disease or disorder.

The FTC's position on antibacterial hand lotion is unambiguous. Jodie Bernstein, director of the FTC's Bureau of Consumer Protection, said in announcing the Unilever settlement, "No marketing claims will wash without adequate substantiation. That's the law. And right now, there's simply no evidence that using antibacterial lotion reduces the incidence of any disease. We want consumers to know that washing well with soap and water is as good as it gets when it comes to warding off disease-causing germs and protecting themselves from colds and flu."

The label on Vaseline Brand Intensive Care Antibacterial Hand Lotion lotion still says "antibacterial," since triclosan is still in there, but with the FTC having enforced all of the above, it would seem difficult for any company to make any claims at all about the beneficial effect in battling infection of any soap or hand lotion containing antibacterials.

Difficult, but not impossible. Ad copywriters have devised many ways of obeying the law while at the same time promoting good personal hygiene and implying that if you don't use their product you will be proving yourself at best somewhat cavalier and irresponsible, and at worst a smelly, germ-infested social outcast. The company that makes Suave Antibacterial Hand Sanitizer, for example, has a helpful "Frequently Asked Questions" section on their web page. "Do hand sanitizers work?" they ask. And they answer their own question, "Yes, using a hand sanitizer can help kill germs you come in contact with every day. To help reduce the number of germs on skin, carry a small tube in your purse or backpack to use when soap and water are not available." Notice that they say that it will reduce the number of germs on skin, but they don't claim that this will improve your health any more than ordinary hand washing—because it won't. Then they say it's good to use it when you don't have any soap and water. In other words, it's better than nothing, but no better than ordinary soap. So, yes, antibacterial hand soaps "work," but it is up to you to decide exactly how well. The active ingredient in Suave Antibacterial Hand Sanitizer is ethyl alcohol, a substance that will unquestionably kill certain bacteria on contact.

In addition to its antibacterial hand sanitizer, Suave also makes an antibacterial cake soap. It's called Suave Antibacterial Deodorant Soap, and its antibacterial ingredient is triclosan. It also contains the following "inactive ingredients": sodium tallowate, sodium cocoate or palm kernelate, water, fragrance, propylene glycol, sodium chloride, glycerin,

disodium phosphate, trisodium EDTA, trisodium etidronate, BHT, FD&C red #4, D&C yellow #8, and D&C yellow #5. Suave also makes a bar soap called Suave Refreshing Deodorant Soap. This one has every single inactive ingredient contained in the Antibacterial Deodorant Soap (except that it has green dye instead of yellow and red). The only thing it lacks is the triclosan. Suave sells the two different kinds of soap over the Internet for the same price: $2.50 for six bars. You get the triclosan for nothing, and it's probably worth it.

There are several peer-reviewed studies of hand sanitizers, most of them carried out by employees of companies that manufacture the stuff. At least two studies show that using antibacterial hand sanitizers works better than not washing at all, but none I have been able to find that shows that antibacterial hand sanitizers are any more effective than ordinary hand washing. One study shows that washing for 20 seconds with ordinary soap and water is more effective than a 70 percent ethyl alcohol hand sanitizer. Dial makes a hand sanitizer that is 62 percent ethyl alcohol, and GOJO Industries makes a hand sanitizer called Purell that is also 62 percent ethyl alcohol. Like triclosan, alcohol has no persistent germ-killing effect. Once it evaporates, its effect stops. Unlike triclosan, alcohol does not encourage the development of resistant strains of bacteria.

Ethyl alcohol sometimes presents a problem: it irritates the skin. In fact, that's one of the warnings printed on the labels of alcohol-based hand sanitizers—if redness or sores appear on your skin, the label says, you should discontinue use and consult a physician. The warning also says that the product will irritate your eyes, so you should keep it away from your face, and if you do get it in your eyes, you should wash them out immediately with clean water and get medical attention if the irritation persists. It also says that you should keep it away from flames, since alcohol is quite flammable. The label ends with a warning to prevent children from getting near it, something you might have already concluded after having read about its other qualities. One would think that most people would hesitate to use a hand-sanitizing product if told that it might irritate their skin and eyes to a point where they would have to call a physician, and that in addition it might under certain circumstances burst into flame.

There are many antibacterial products designed for cleaning kitchens, bathrooms, and other parts of the house. The Clorox Company is a large producer of such products. Their cleaning products contain a

very effective and very inexpensive disinfectant, chlorine bleach. Clorox Clean-Up is recommended for kitchens and bathrooms. Clorox Automatic Toilet Bowl Cleaner is designed for toilet bowls ("Just drop the tablet in your toilet tank, and with every flush, it will clean, deodorize and kill 99.9% of germs in the bowl water"). It also comes in a gel form that "sticks to the sides of the bowl while it cleans tough stains to leave the bowl sparkling clean and sanitary." For laundry use, the makers recommend Clorox® Liquid Bleach, whose active ingredient, like all of the Clorox products, is sodium hypochlorite. Notice the strategic placement of the ® in the name of their liquid bleach—only the "Clorox" is registered. Sodium hypochlorite, or liquid bleach, can be sold by anyone and registered by no one—the sodium hypochlorite sold under the supermarket house brand name is no different from the chemical Clorox puts in its antibacterial cleaning products.

Chlorine bleach is, without doubt, an effective germ killer. It will even kill viruses, which is why it is used for cleaning up blood and body fluids that are potentially infected with HIV. The Clorox Company recommends it for cleaning your clothes; cleaning up a toilet backup; purifying drinking water; cleaning combs, brushes, coffeemakers, flowerpots and planters; sanitizing the kitchen sink, sponges, countertops, cutting boards, and dishcloths; cleaning the kitchen floor, stovetop, and trash cans; spraying on surfaces inside the refrigerator; and cleaning the tub, sink, toilet bowl, and floor in the bathroom. They suggest sprinkling it in the kids' sandbox and wading pool; using it to remove mold from steps, siding, tile, bricks, stucco, and stone; and applying it to any number of things that little kids touch in order to protect them from germs. If you did everything Clorox tells you to do with chlorine bleach, it seems, you'd have little time to do anything else and everything in and outside your house would smell like a swimming pool. Of course, Clorox doesn't specify exactly what good all of this will do for your health, since it probably will do little or nothing that routinely conscientious housecleaning wouldn't do just as well.

The EPA, For Your Information

An antibacterial impregnated in a toy or in some other consumer product must be registered with the Environmental Protection Agency. Companies that manufacture such things have correctly perceived that

germs on cutting boards and in children's toys are among those people worry most about, and they've come to the rescue with antibacterial products they claim will protect you from all sorts of diseases. Unfortunately, there aren't any scientific data to support a belief that products impregnated with antibacterials will protect you from anything.

The EPA doesn't like it when manufacturers make antibacterial claims about their products that are unsupported by scientific fact, and they are in a position to do something about it. In January 1998, the EPA ordered seven companies to stop making unlawful claims that their products protected people against bacteria. The products involved were cutting boards, kitchen gadgets, and utensils that the companies claimed were impregnated with antibacterials in such a way as to provide public health protection. The companies included Arrow Plastics Manufacturing Co.; Acrylic Plastics Products, Inc.; Farberware Lifetime Hoan; Bradshaw International Inc.; Decorator House Inc.; Better House Corp.; and Pacific International Group. If you own antibacterial kitchen equipment manufactured by any of these companies, you should treat it the same way you treat any other kitchen equipment, because none of these products prevent the growth of bacteria that cause food-borne illnesses—even though their labels claimed they would.

It's legal to manufacture products that have antibacterials embedded in them, but it's not legal to claim they protect their users against disease. As an example of an acceptable label on such a product, the EPA suggests the following: "Antibacterial properties are built in to inhibit the growth of bacteria that may affect this product. The antibacterial properties do not protect user or others against bacteria, viruses, germs, or other disease organisms. Always clean and wash this product thoroughly before and after each use." Even the name of the product can't imply that it will make you healthier by killing germs—to use a made-up example, a name like "Health Defender Antibacterial Cutting Board" would be unlikely to pass muster with the EPA.

Hasbro, the company that made Playskool toys embedded with Microban, found themselves in a bit of trouble, too. No sooner had they started selling their new line of antibacterial toys than the EPA came down on them. In April 1997, Hasbro reached an agreement with the EPA to stop claiming that toys treated with Microban would protect children from infectious diseases caused by bacteria. Not only did Hasbro agree to stop making those claims, but they also agreed to run

advertisements in magazines and newspapers and in toy stores where their products are sold correcting any wrong impression they might have created. On top of that, they paid a $120,000 fine. All told, their false claims about Microban in their toys cost them about a million bucks. This was apparently enough to convince them that it wasn't really worth the trouble. Hasbro no longer manufactures or sells Microban-embedded toys.

There are still plenty of products embedded with Microban, if you want to buy them. Microban Products Company still sells the substance and the process for embedding it in products. They are not fools, however, and they know just how much benefit they're allowed to claim for their product. Their advertising is conspicuously lacking in any health promises—it's the product itself that is protected from bacteria, not the person who uses it. Even claims about the protection the stuff confers on the product have their limits—limits established by various regulating agencies, not by Microban. Microban explicitly states, for example, that you still have to clean things embedded with their product. "Absolutely," they say on their web site. "Good hygiene practices should always be observed, including the washing of hands and cleaning of products. Microban active ingredients provide added protection in-between cleanings." Added protection from what? They don't say. They do helpfully add that their products have lots of regulatory approval: "Microban Products Company has registered active ingredients with the US Environmental Protection Agency (EPA)," they boast, "And products with Microban Built-In Protection have been approved by the US Food & Drug Administration (FDA), the European Union (EU), the National Sanitation Foundation (NSF) and other regulatory authorities around the world." They don't add that without those approvals, they wouldn't be allowed to sell the stuff.

The benefits for product manufacturers of embedding the things they sell with antibacterials are spelled out by the Microban Products Company: " . . . [T]he benefit is a real differentiating advantage versus competitors that is both compelling to end-users and a driver of profitable growth." The benefit to you is somewhat less specific and not nearly so compelling: "For the end-user, the benefits of the Microban brand promise are peace of mind and increased cleanliness."

I have been pretty hard in this chapter on companies that sell antibacterial household products, but it must be remembered that they are

responding to a correctly perceived consumer demand for such things. When I spoke to Procter & Gamble to ask if they made an antibacterial hand soap or sanitizer, they could have said, "No, we don't, but if you wash your hands with Ivory Soap, which we do make, you'll get the same germ-fighting effect as you will with antibacterial hand sanitizers, maybe even a little better." If they had said that, they would have told the truth and maybe even sold a couple more bars of soap. Instead, they explained, trying quite honestly to be helpful, that no, we don't make an antibacterial hand soap, but we have a dish soap called Dawn that's antibacterial, and that can also be used as a hand soap. Dawn has triclosan in it.

Should you buy and use antibacterial soap, laundry detergent, cleaning products, and the hundreds of other antibacterial products now out there for sale? These days it may be difficult to find a cleaning product that *isn't* labeled "antimicrobial" or "antibacterial," so sometimes you're not going to have any choice. For example, one survey found that on supermarket shelves 76 percent of liquid hand soaps and 29 percent of bar soaps contained antibacterial agents, mainly triclosan. But if you consider the known benefits of such concoctions and their known risks, the balance weighs very heavily in favor of avoiding them. There is no evidence that they are any better for your health than ordinary products; there is considerable evidence that at least some of them can be harmful. This being the case, it is difficult to see why anyone would go out of his way to use them.

Glossary

acanthamoeba An amoeba that lives in water and moist soil and can cause a fatal inflammation of the brain, meningoencephalitis.

adenovirus A virus of the family adenoviridae, which cause a large variety of human illnesses.

Aedes triseriatus The species of mosquito that transmits La Crosse encephalitis.

Aeromonas A gram-negative rod-shaped bacillus. One subspecies, *A. hydrophila*, lives in fresh water and causes cellulitis on skin lacerations.

Alphavirus A genus of viruses transmitted by mosquitoes. They cause several different kinds of encephalitis.

amantadine An antiviral medicine used in preventing lung infections from flu viruses.

amoxicillin/clavulanate An antibiotic used in treating pneumonia and tuberculosis.

amphotericin B An antifungal drug used in treating candidiasis among other disorders.

ampicillin One of the beta lactam antibiotics closely related to penicillin.

anaerobic Existing in the absence of oxygen.

Ancylostoma braziliense A nematode parasite carried by dogs that causes skin disease in humans.

antigenic drift A gradual change in surface antigens of a microorganism by which it can acquire resistance to an antimicrobial medicine.

antigenic shift An abrupt change in surface antigens of a microorganism by which it can acquire resistance to an antimicrobial medicine.

arboviruses A recent coinage that stands for arthropod-borne viruses. These are viruses carried by arthropods, invertebrate animals that have segmented external coverings and jointed limbs.

archaea Sometimes called archaeobacteria. A group of unicellular organisms

struturally different from bacteria and viruses. They often inhabit environments chemically or thermically fatal to most forms of life.

arthritis Inflammation of the joints.

Aspergillus A genus of fungi that includes several species that cause human illness. Subspecies include *A. flavus, A. fumigatus,* and *A. versicolor.* They cause infections of the lungs and bronchi, usually in debilitated patients.

azithromycin A macrolide antibiotic (similar to erythromycin) useful in treating chlamydia infections.

bacillus A rod-shaped bacterium. The plural is bacilli.

Bacillus cereus A bacterium that has two forms, one causing vomiting, the other diarrhea. It is often implicated in cases of food poisoning, especially contaminated rice. Even fried rice can be infected—the bacillus is fairly stable when heated.

bacteria Plural of bacterium. A general term for single-celled microorganisms, including many that cause human disease. They are classified according to shape: rod-shaped bacilli, comma-shaped vibrios, spherical cocci, and spiral-shaped spirilla.

bacteriophage Also called a phage; the simplest form of reproducing life known. It consists of a nucleic acid core and a protein coat. It can destroy bacterial cells under certain conditions.

Bactrim Brand name of a combination of trimethroprim and sulfamethoxazole, two sulfa drugs used in treating various urinary tract infections, ear infections caused by *Streptococcus pneumoniae* or *Haemophilus influenzae,* shigellosis, *Pneumocystis carinii* pneumonia, and travelers' diarrhea caused by *E. coli.* It is also used for treating infections of *Cyclospora cayetanensis.* Other brand names are Cotrim and Septra.

Bartonella A species of small gram-negative bacillus whose subspecies include *B. clarridgeiae, B. henselae, B. koehlerae, B. quintana,* and *B.bacilliformis.* These organisms cause a wide variety of human diseases, including cat-scratch disease caused by *B. henselae.*

beta lactam antibiotics A group of antibiotics characterized by their similar molecular structure that includes penicillin, amoxicillin, and ampicillin among other agents. Sometimes styled ß-lactam.

bichloroacetic acid (BCA) A topical treatment for genital warts.

Bordetella pertussis The gram-negative bacillus that causes whooping cough.

Borrelia burgdorferi The bacterium that causes Lyme disease.

bradyzoite One of three stages in the development of certain protozoan parasites.

bronchiolitis Inflammation of the bronchioles, small tubes in the lungs.

Campylobacter A genus of spiral gram-negative bacteria that live in the intestinal tract and cause infectious diarrhea and inflammation of the stomach. There are several subspecies including *C. jejuni* and *C. intestinalis*.

Candida albicans A type of yeast-like fungus. In humans it causes thrush, vaginitis, and sometimes systemic infections.

capsid The protein coat of a virus.

cefazolin A cephalosporin antibiotic useful in treating a variety of bacterial illnesses including *Streptococcus* infections, *Haemophilus influenzae*, *Staphylococcus* infections, and *E. coli* infections, among others.

cefixime A cephalosporin antibiotic indicated for treating urinary tract infections and ear infections.

ceftriaxone A broad-spectrum cephalosporin antibiotic effective against many different bacteria, including *Streptococcus, Staphylococcus, Haemophilus influenzae*, and others.

cellulitis Acute painful inflammation of the skin.

cephalosporins A large group of antibiotics that are effective against gram-positive and some gram-negative bacteria.

Cephalosporium A type of mold common in houses.

cerebrospinal fluid The fluid surrounding the brain and spinal cord. Examination of the cerebrospinal fluid is useful in diagnosing various illnesses.

chancre The lesion caused at the site of infection by syphilis. It is a hard, reddish sore with a yellowish secretion.

Chlamydia A genus of gram-negative bacteria. *C. psittaci* is the cause of psittacosis; *C. trachomatis* causes conjunctivitis, urethritis, and proctitis.

chloramphenicol A broad-spectrum antibiotic originally obtained from a plant but now produced synthetically. While it is effective against many strains of both gram-positive and gram-negative bacteria, it is rarely used because of a rare but very serious side effect, a blood disorder called aplastic anemia.

cholera An acute infectious disease caused by *Vibrio cholerae* and characterized by severe diarrhea. It is transmitted by the oral-fecal route in infected water supplies, and is epidemic in Asia but almost nonexistent in the United States.

ciguera toxin (ciguatoxin) Toxin produced by the dinoflagellate *Gambierdiscus toxicus*.

Ciprofloxacin A synthetic broad-spectrum antimicrobial agent effective against dozens of different gram-positive and gram-negative microorganisms, including gonorrhea, *Salmonella, Shigella*, and *Enterococcus faecalis*, among many others. It is also effective in treating anthrax.

Cladosporium herbarum A fungus commonly found in houses.

Clostridium botulinum The gram-positive anaerobic bacterium that causes botulism.

Clostridium difficile A bacterium of the same genus as *C. botulinum* that causes diarrhea; typically nosocomially transmitted.

coccus A round or oval-shaped bacterium. Plural is cocci.

corneal lesions A morbid change in the cornea, the transparent part of the outer coat of the eyeball.

coronavirus Single-stranded RNA viruses whose name derives from the club-shaped arms on the viral envelope. They are among the hundreds of viruses that cause colds.

Corynebacterium A genus of gram-positive club-shaped bacteria, one of whose species, *C. diphtheriae*, causes diphtheria.

Cotrim See **Bactrim.**

Coxsackie virus One of the enteroviruses. Can harm the brain, heart, respiratory tract, or skin. Named after Coxsackie, New York, where it was discovered.

Creutzfeldt-Jakob disease A neurological disease that is usually genetic, but that has a variant form now linked to the prion that causes bovine spongiform encephalopathy. It causes progressive dementia and is invariably fatal.

cryptosporidiosis A diarrheal disease caused by *Cryptosporidium parvum*. It is usually mild, but can be severe in immunocompromised patients.

Cryptosporidium parvum A waterborne parasitic protozoan that causes diarrhea when ingested.

Cyclospora cayetanensis A unicellular parasite that is food borne and causes diarrhea.

cutanea larva migrans A skin infection caused by, among other things, the nematode *Ancylostoma braziliense*, or hookworm.

DEET (n,n-diethyl-m toluamide) An insecticide useful in protecting against the ticks that cause Lyme disease.

dinoflagellates A group of more than 2,000 different species of marine organisms characterized by a cellulose envelope with two flagellae used for locomotion.

diphtheria An acute contagious disease, vaccine preventable, caused by a bacillus, *Corynebacterium diphtheriae*.

DNA Dioxyribonucleic acid, the molecular basis of heredity, responsible for the replication of proteins and nucleic acid.

DOTS Directly observed treatment—short term. A method of treating tuberculosis that involves watching a patient to make sure that the prescribed medicine is ingested.

doxycycline A synthetic broad-spectrum antibiotic useful against many microorganisms.

E. coli O157:H7 A deadly strain of the bacterium *Escherichia coli.*

Eastern equine encephalitis (EEE) A rare but sometimes fatal arbovirus caused by viruses of the *Togaviridae* family and carried by mosquitoes.

Ebola A virus of the family *Filoviridae* (Marburg is another member of this family) that causes a systemic disease with a high mortality rate. It first appeared in 1976 in Zaire and Sudan.

ELISA Enzyme-linked immunosorbent assay, a laboratory test used as an initial screen for HIV and other viruses.

emphysema A swelling caused by abnormal presence of air, usually referring to a swelling of the lungs, in which case it is called pulmonary emphysema.

encephalopathy Disease of the brain.

enterobacter cloacae A (usually) nosocomial infection that can cause pneumonia, urinary tract infections, and brain abscesses.

enterococci Any *Streptococcus* that lives in the intestines.

enterotoxin A poison that specifically attacks the cells of the mucous membrane of the intestines.

epidermis The outer layer of the skin.

epididymis A structure in which sperm are stored. It is attached to the testes at one end and the deferent duct at the other.

Escherichia coli A gram-negative bacterium that is part of the normal flora of the human intestines. Some species are pathogenic.

ethambutol A compound active against the mycobacterium that causes tuberculosis.

facultative Refers to microorganisms able to live under more than one set of environmental conditions.

fatal familial insomnia A rare hereditary disease that destroys the part of the brain called the thalamus, eventually causing death. A nonhereditary variant of it is caused by a prion similar to the one that causes Creutzfeldt-Jakob disease.

flagellae Whiplike protein structures used by some microorganisms for locomotion.

Flavivirus A genus of arthropod-borne viruses including the ones that cause yellow fever, dengue, and encephalitis.

fluorescent-antibody testing A diagnostic test for the presence of antibodies to microorganisms.

fluoroquinolones A group of antibiotics effective against most gram-negative rods. Ciprofloxin and ofloxacin are among them.

fungus A member of the phylum Fungi, characterized by asexual reproduction and lack of chlorophyll. Fungi are usually parasitic.

Fusarium A genus of fungi whose species are commonly poisonous to other plants and to humans, affecting the circulatory, digestive, skin, and nervous systems.

genus-specific fluorescein-tagged monoclonal antibody staining A technique in pathology of staining microorganisms to identify them by determining the pH of the cell.

Gerstmann-Straussler-Scheinker syndrome A rare degenerative nerve disorder, usually hereditary, but can be caused by a prion.

Giardia A genus of protozoa, some of whose species parasitize the intestines of humans. *Giardia lamblia* is a common cause of diarrhea and other intestinal symptoms.

gonorrhea A sexually transmitted disease caused by the gram-negative bacterium *Neisseria gonorrhoeae*.

gram-negative Descriptive of microorganisms that cannot be stained by the violet dye of Gram's stain.

gram-positive Descriptive of microorganisms that can be stained by the violet dye of Gram's stain.

granulomatous amoebic encephalitis (GAE) A brain disorder caused by the parasitic *Acanthamoeba*.

Guillain-Barré syndrome A neurological disease that involves the degeneration of the myelin coating of the nerve cells. Sometimes appears after a minor febrile illness.

Haemophilus influenza **type b (Hib)** A vaccine-preventable cause of bacterial pneumonia, particularly in children. The pathogenic agent is a strain of the *Haemophilus influenzae* bacterium.

hantavirus A deadly viral disease (mortality rate about 40% with the best treatment) caused by a rodent-borne virus discovered in 1993.

Helicobacter pylori The bacterium that is the major cause of peptic ulcer in humans.

hemagglutinin A blood protein that causes red blood cells to clump together. It is also present in some viruses in which it allows the virus to infect a cell.

hemolytic uremic syndrome (HUS) A potentially lethal blood syndrome that occurs most often in children, and is sometimes preceded by respiratory or intestinal infection.

hepatitis A One of the viruses that causes viral hepatitis. It is vaccine-preventable.

hepatitis B A virus that causes one form of viral hepatitis. It is vaccine-preventable.

hepatitis C A virus that causes a chronic form of liver disease.

hepatitis Inflammation of the liver. Usually refers to a liver disease caused by one of several different viruses.

Herpes simplex I The strain of herpes virus that causes cold sores.

Herpes simplex **II** The strain of herpes that affects the genital organs and is sexually transmitted.

HIV Human immunodeficiency virus, the virus that causes acquired immune deficiency syndrome (AIDS).

human immunodeficiency virus See **HIV**.

human papillomavirus (HPV) A genus of viruses some of whose species cause warts.

human parainfluenza virus (HPIV) A type of virus usually acquired in childhood so that most children have antibodies to it by the age of eight years. While HPIV is a common cause of respiratory illness in children, it is rare in adults, and almost always causes only mild illness.

hypoxia Oxygen deficiency.

interferon A protein produced by the body to fight virus infection. Also, a synthetic drug used to treat viral diseases.

intussusception A twisting of the intestine that causes acute abdominal symptoms, usually in children.

isoniazid A drug, isonicotinic acid dydrazide, used in the treatment of tuberculosis.

Ixodes A genus of ticks that transmit diseases to humans, including Lyme disease, babesiosis, human granulocytic ehrlichiosis, Rocky Mountain spotted fever, and others.

Klebsiella A genus of gram-negative bacteria that includes *K. pneumoniae*, which causes a kind of bacterial pneumonia and other infections. *K. oxycytoca* is a recently discovered species. These are common nosocomial infections and are notoriously resistant to antibiotics.

kuru A degenerative neurological disease caused by a prion and transmitted by ingestion of the brains of people who have died of it.

La Crosse encephalitis A rare disease of the central nervous system transmitted to humans by mosquitoes. There are about 70 cases per year.

Legionnaire's disease A pneumonia caused by a gram-negative rod-shaped bacterium, *Legionella pneumophila*.

leptospirosis Infection with the bacterium *leptospirum*, which is carried by rats and other animals. Can cause symptoms of varying severity from mild fever to jaundice to renal failure.

leukocidin A toxin that destroys the leukocytes in human blood.

lipids The fats that along with proteins and carbohydrates compose the major parts of the cell.

Listeria monocytogenes A gram-positive bacterium that causes listeriosis, characterized by meningitis and septicemia. It belongs to the family *Corynebacteria*.

lumbar puncture A surgical procedure to sample the spinal fluid for diagnostic purposes.

Lyme disease A disorder caused by a tick of the *Ixodes* genus, characterized by muscle pain, arthritis of the joints, and in some cases nervous system involvement.

LYMErix A vaccine effective against Lyme disease.

lysozyme Mucopeptide glucohydrolase, an enzyme secreted in tears, sweat, saliva, and nasal discharge that functions as a natural antibiotic to fight infection.

macrolide antibiotic A group of antibiotics of a particular molecular structure. Includes erythromycin.

malaria An infectious disease caused by a parasite and transmitted to humans by mosquitoes of four different species.

Marburg A virus of the family *Filoviridae* (the same family as Ebola) that causes an acute febrile illness with a high fatality rate. It was first isolated in 1967 in laboratory workers who had been exposed to African green monkeys.

measles A highly contagious, vaccine-preventable viral disease characterized by fever and inflammation of the skin and respiratory tract.

meningitis Inflammation of the meninges, the membranes that cover the brain and spinal cord.

meningococcal meningitis A form of meningitis caused by the gram-negative bacterium *Neisseria meningitidis*. There is a vaccine that prevents it.

Menomune Brand name of the meningitis vaccine manufactured by Aventis Pasteur.

methicillin-resistant *Staphylococcus aureus* (MRSA) A bacterial infection resistant to the antibiotic methicillin. A dangerous form of staph infection.

metronidazole A drug used in the treatment of rosacea, an inflammatory skin disorder. Also used to treat trichomoniasis infection.

metronidazole An antiparasitic drug used in the treatment of giardiasis and trichomoniasis.

microimmunofluorescence assay A complex and highly sensitive laboratory test designed to identify microorganisms.

mumps A highly contagious vaccine-preventable disease caused by an RNA virus and characterized by swollen glands and fever. It can also involve other organs and the central nervous system.

Mycobacterium A genus of gram-positive rod-shaped bacteria that includes the bacteria that cause tuberculosis, leprosy, and other diseases.

Mycobacterium fortuitum A species of the genus *Mycobacterium* that is waterborne and causes skin rashes and ulcers in humans.

Mycobacterium tuberculosis The microorganism that causes tuberculosis.

mycotoxin A poison produced by a fungus.

myiasis A disease caused by insects or insect larvae that harbor in living tissue.

myocarditis Inflammation of the muscle of the heart.

mysophobe A person irrationally afraid of germs.

Naegleria fowleri A ubiquitous, free-living amoeba that parasitizes humans and causes encephalitis by penetrating the brain through the nasal passages. Most infections are fatal. Infections of the eye lead to blindness.

Neisseria gonorrheae See **gonorrhea.**

Neisseria meningitidis See **meningococcal meningitis.**

nephritis Inflammation of the kidneys.

neuraminidase An enzyme required by viruses for releasing progeny from infected cells. Certain antiviral drugs function by inhibiting the production of this enzyme.

neurosyphilis Syphilis that has infected the nervous system.

neurotoxin A substance poisonous to nerve cells.

Norwalk-like virus (NLV) One of the caliciviruses. The symptoms of infection with this organism can include nausea, vomiting, diarrhea, and abdominal cramps. Often associated with eating raw oysters.

nosocomial infection An infection usually confined to or transmitted in hospitals.

nucleic acid hybridization and sequencing A laboratory test for identifying microorganisms by examination of their DNA.

nucleocapsid The nucleic acid of a virus enclosed by its protein capsid, or coat.

obligate Able to survive in only one kind of environment. The opposite of facultative.

ocular larva migrans The organism responsible for toxocariasis, a usually mild but sometimes severe disease that can scar the retina and cause blindness. The germ is often found in dog stool.

oocyst The encysted stage of the reproductive cycle in which the formation of sporozoites, the infectious stage of sporozoans, occurs.

oseltamivir A recently developed medicine for preventing and treating influenza types A and B. Sold under the brand name Tamiflu®.

otitis media Inflammation of the middle ear. Often one of the consequences of respiratory infection.

palivizumab (Synagis) A monoclonal antibody produced by recombinant DNA technology for the prevention of respiratory syncitial virus in infants.

Pap smear A test in which cells are examined for abnormal growth indicating cancer. Used especially for detection of cancer in the female genital tract and for the evaluation of hormonal status. Named after its developer, George Nicolas Papanicolaou.

Papillomavirus A genus of virus that includes the ones that cause warts in humans.

pelvic inflammatory disease (PID) Inflammation of the uterus and/or fallopian tubes, caused either by medical procedures or pathogens.

PCR Polymerase chain reaction. A technique for amplifying viruses for detection and identification.

Penicillium A genus of fungi that produces several different antibiotic substances.

Penicllium chrysogenum A species of penicillium that produces penicillin.

Penicillium expansum A species of penicillium that causes fruit decay and produces the carcinogenic mycotoxin patulin.

Penicillium notatum A fungus from which penicillin was first extracted.

perchloroethylene A solvent used in dry cleaning.

pertussis Whooping cough. An acute respiratory illness, usually in infants and children, caused by the bacterium *Bordatella pertussis*, a gram-nega-

tive coccobacillus. It is characterized by violent coughing, and is vaccine-preventable.

phagocyte A cell that fights invading foreign particles by ingesting them.

picornavirus A large family of viruses that includes species that cause, among other disorders, polio, hepatitis A, meningitis, and the common cold.

Plasmodium The family of protozoan parasites that includes the four species that cause different types of malaria: *P. falciparum, P. malariae, P. ovale,* and *P. vivax.*

pleconaril A drug under development for the treatment of colds.

Plesiomonas shigelloides An anaerobic gram-negative rod-shaped bacterium that causes intestinal illness in humans, mostly in the immuno-compromised, and is transmitted in infected water or food.

pneumococcal pneumonia An acute pneumonia caused by *Pneumococcus.*

pneumonitis Inflammation of the lungs.

podophyllin resin A bitter-tasting caustic substance produced by the roots of May apples and used as a cathartic and as a topical treatment for warts.

Porphyromonas gingivalis A gram-negative anaerobic bacterium that causes gum disease and tooth loss.

Powassan encephalitis One of several viral diseases transmitted by *Ixodid* ticks.

prions Abnormal cellular proteins that can spread from cell to cell. They cause Creutzfeldt-Jakob disease and other degenerative nerve disorders.

Propionibacterium acnes The gram-positive anaerobic rod-shaped bacterium that lives in human sebaceous glands and causes acne.

protozoan A one-celled organism of the phylum *Protozoa.*

Pseudomonas A gram-negative flagellated bacterium found in soil, water, sewage, and the air. It causes infections in wounds and burns.

purified-surface-antigen preparation A method of vaccine production that uses only the part of the virus that provokes an immune response instead of using the whole virus.

pyrazinamide One of the several different medications used in the treatment of tuberculosis.

pyrimethamine An antiparasitic compound used in the treatment of malaria and toxoplasmosis.

rabies A viral encephalitis transmitted to humans by the bite of infected animals.

rat-bite fever A disease transmitted by rats, mice, gerbils, and other small rodents and animals that eat them. It is caused by *Streptobacillus moniliformis,* a gram-negative rod-shaped bacterium.

rattus norvegicus Norwegian rat. This is the species most common in the United States.

respiratory syncytial virus (RSV) The major respiratory illness in young children and the principle cause of lower respiratory disease in infants. The pathogen is a member of the **Pneumovirus** genus.

reverse transcriptase-polymerase chain reaction (RT-PCR) A complex laboratory test for the presence of viral RNA.

rhinovirus A genus of the family *Picornaviridae* that causes colds in humans and foot-and-mouth disease in cattle.

rifampin A drug used in combination with other drugs to treat methicillin-resistant *Staphylococcus* infections. Also used as prophylactic treatment for Legionnaire's disease and meningococcal meningitis and, in combination with other drugs, to treat tuberculosis.

rimantadine An antiviral effective against influenza A.

RNA Ribonucleic acid. A family of polynucleotides, components of all living cells. There are various kinds of RNA including messenger RNA, ribosomal RNA, and transfer RNA, each of which has a different function in the cell's physiology.

Rotashield A vaccine for rotavirus, approved for use and then quickly taken off the market when its dangerous side effects were discovered.

Rotavirus A genus of the family *Rotaviridae*. It is the most common cause of severe dehydrating diarrhea in children under three years old.

rubella German measles. A contagious disease, usually of short duration, that can cause birth defects when contracted by pregnant women.

Salmonella A genus of gram-negative bacteria, several of whose species, including mostly commonly *S. enteriditis* and *S. tymphimurium*, cause intestinal illness.

salmonellosis The intestinal illness caused by *Salmonella*.

scabies A skin disorder caused by a mite that deposits eggs under the skin of humans, causing skin irritation.

scombroid poisoning A histamine nerve poisoning caused by a toxin produced by dinoflagellates that parasitize fish of the scombroid family, including tuna, mackerel, and dolphin fish.

septicemia A systemic illness caused by microorganisms of many different kinds.

Septra See **Bactrim.**

serum Specifically, the clear fluid of blood. Often used as a general term for the parts of blood that contain antitoxins that can be analyzed in the laboratory for diagnostic purposes.

Shigella A genus of gram-negative bacteria that causes dysentery. Includes *S. dysenteriae* (Group A shigella), *S. flexneri* (Group B Shigella), *S. boydii* (Group C Shigella), and *S. sonnei* (Group D shigella), all of which cause infections of varying degrees of severity. These are very closely related to *E. coli.*

shigellosis The intestinal illness caused by the *Shigella* bacteria.

shingles A painful skin infection caused by *varicella zoster.*

smallpox A contagious viral illness characterized by facial and body pustules and scarring. It is also called variola, and it is the only disease ever wiped out by man.

spiramycin A drug used in the treatment of toxoplasmosis, particularly in pregnant women.

spirilla Spiral-shaped bacteria.

sporozoite A stage in the life cycle of sporozoan organisms. In malaria, it is this stage that causes illness.

St. Louis encephalitis One of several viral diseases transmitted to humans by ticks of the *Ixodes* genus.

Stachybotrys chartarum A common type of household mold.

Staphylococcus aureus A species of the gram-positive genus *Staphylococcus* and the cause of many different infections and disorders in humans.

Staphylococcus epidermidis A species of the gram-positive genus *Staphylococcus* that is part of the normal flora of the human skin.

Staphylococcus hyicus A species of the gram-positive genus *Staphylococcus* that has been implicated in food poisoning.

Staphylococcus intermedius A species of the gram-positive genus *Staphylococcus* sometimes associated with skin infections caused by dog bites.

Streptobacillus moniliformis A gram-negative rod-shaped bacterium that causes rat-bite fever.

Streptococcus fecalis A gram-positive round bacterium that is a normal part of the intestinal flora and that can infect the blood, causing endocarditis, an infection of the membrane that lines the heart.

Streptococcus mutans A gram-positive species of the genus *Streptococcus* that is one of the constituents of dental plaque.

Streptococcus pneumoniae (*pneumococcus*) A gram-positive species of the genus *Streptococcus* that causes pneumonia, middle ear infections, and meningitis.

Streptococcus sanguis A gram-positive species of the genus *Streptococcus* that is one of the constituents of dental plaque.

streptomycin The first drug used to treat tuberculosis. Now tuberculosis is largely resistant to it.

subvirion preparation A type of vaccine prepared by treating a virus chemically so that it is broken down into smaller subunits called subvirions. Subvirions provoke the immune response without exposure to the disease-causing virus.

sulfadiazine A drug used in the treatment of toxoplasmosis in pregnant women.

sulfonamides A general term for the antibiotic sulfa drugs.

syphilis A sexually transmitted disease caused by the bacterium *Treponema pallidum*.

tachyzoite A stage in the life cycle of certain protozoa. In toxoplasmosis, it is this state that causes the acute infection.

tetanus A central nervous system disease caused by *Clostridium tetani*, a gram-positive anaerobic bacterium.

tetracycline A broad-spectrum antibiotic.

T-helper 1 (TH-1) cells The blood cells that produce cell-mediated immunity by binding to the surface of cells that display an antigen.

T-helper 2 (TH-2) cells The blood cells that produce antibody-mediated immunity by provoking the production of antibodies that bind to antigens.

thiabendazole A drug for treating cutaneous larva migrans, the result of hookworm infection.

thrombocytopenia A condition characterized by an abnormally low concentration of platelets in the blood.

Togaviridae A family of viruses that includes those that cause yellow fever and German measles.

toxocariasis Infection with *Toxocara*, a parasitic roundworm.

Toxocaris canis The species of *Toxocara* that is transmitted from dogs to humans.

Toxoplasma gondii The protozoan parasite that invades cells and causes toxoplasmosis in humans. Often transmitted by cat feces.

toxoplasmosis The disease caused by *Toxoplasma gondii*, which is usually mild but can be spread in a more serious form by a pregnant mother to her child.

Treponema pallidum The spiral-shaped bacterium that causes syphilis.

Treponema pallidum pertenue The subspecies of *Treponema* that causes yaws.

trichloroacetic acid (TCA) A topical treatment for warts.

Trichoderma Fungi that live in soil and other places, including houses.

trichomoniasis An infection of the vagina or the urinary tract of men, caused by the parasite *Trichonoma vaginalis*.

Trichonoma vaginalis The protozoan parasite that causes trichomoniasis.

triclosan An antibiotic added to hand cleaners, floor cleaners, and many other products, edible and not, to make them antibacterial. It works by interfering with an enzyme the bacterium needs for growth.

trimethoprim-sulphamethoxazole A combination of antibiotics used in the treatment of infections of the urinary tract, shigellosis, and *Pneumocystis* pneumonia.

trophozoite The adult stage of a sporozoan protozoan, the stage in which it feeds.

tuberculosis The infectious disease, commonly of the lungs and bones but of other organs as well, caused by *Mycobacterium tuberculosis*.

tumorigenicity The degree to which a substance or disease causes tumors.

typhoid fever A systemic illness caused by the bacterium *Salmonella typhi*, often spread in food contaminated with feces. There are about 500 cases a year in the United States.

typhus A louse- or tick-borne group of illnesses caused by various species of *Rickettsia* bacteria. There are several types, each caused by a different species of *Rickettsia*, including epidemic typhus, Brill-Zinsser disease, murine typhus, scrub typhus, and Rocky Mountain spotted fever.

vancomycin An antibiotic active against all gram-positive cocci and bacilli that works by interfering with the germ's ability to create a cell wall. It is often the antibiotic of last resort, although there are now vancomycin-resistant strains of bacteria. The drug must be given intravenously.

Varicella zoster The virus that causes chicken pox, an acute contagious disease, usually in children. The virus persists in the body after recovery from the initial infection and may reassert itself many years later as shingles.

Vibrio cholerae The bacterium that causes cholera, an infectious disease characterized by severe dehydrating diarrhea.

Vibrio parahaemolyticus A gram-negative bacteria that causes acute abdominal systems in humans. Sometimes found in oysters.

Vibrio vulnificus A species of *Vibrio* virus often implicated in food poisoning.

vibros The group of bacteria that are shaped like commas.

viral hemorrhagic fever (VHF) A group of the most deadly viral diseases, including Ebola, Marburg, dengue, Lassa fever, Machupo virus, and Junin virus, none of which exist in the United States. However, there is one VHF that does exist in this country: hantavirus pulmonary syndrome, a disease caused by a virus carried by deer mice and transmitted to humans when dust from dried mouse droppings is inhaled.

virus An infectious parasite capable of living and reproducing only within other cells and composed of a protein shell that usually encases a nucleic acid, either RNA or DNA.

Western equine encephalitis (WEE) A viral disease caused by a member of the *Alphavirus* genus and transmitted to humans by mosquitoes. WEE occurs all over the country, and disproportionately among infants under one year.

yaws An infection caused by the spirochete *Treponema pallidum pertenue*. Causes skin ulcers, and in advanced cases mutilating facial lesions.

yellow fever A viral infection transmitted by mosquitoes. The illness is endemic in certain parts of Africa, South America, and Central America. It is an unpleasant but usually not serious disease from which people recover without consequence.

Yersinia A genus of rod-shaped gram-negative bacteria, two of whose species cause stomach illnesses in humans.

Yersinia enterocolitica One of the pathogenic species of *Yersinia*. It causes gastroenteritis.

zanamivir A medicine to treat influenza A and B. It is inhaled. The brand name is Relenza.

Notes

Chapter 1. Microbes: An Introduction

Figure 1, and the statistical information in the beginning of this chapter, are from one of the best-edited and most interesting weekly newsmagazines in the United States, the *Morbidity and Mortality Weekly Report*, published by the Centers for Disease Control. This number is entitled "Achievements in Public Health, 1900–1939," July 30, 1999. Alexander Fleming's work has been described in numerous biographies and other works, useful among which is Gwyn Macfarlane *Alexander Fleming: The Man and the Myth* (Cambridge: Harvard University Press, 1984). Perhaps the best source for information on streptomycin is the book by the discoverer himself, Selman A. Waksman, *My Life with the Microbes* (New York: Simon & Schuster, 1954). The article in *Science* concerning anaerobic bacteria that remove the zinc from waste water is M. Labrenz et al., "Formation of sphalerite (ZnS) deposits in natural biofilms of sulfate-reducing bacteria." *Science*. 2000 Dec 1; 290(5497): 1744–47. The bacterium that dechlorinates chlorobenzenes is described in L. Adrian, U. Szewzyk, J. Wecke, and H. Gorisch, "Bacterial dehalorespiration with chlorinated benzenes." *Nature*. 2000 Nov 30; 408(6812): 580–83.

Chapter 2. Hungry? Let's Eat: The Contaminated Kitchen

The account of the rabid cow is from the *Morbidity and Mortality Weekly Report (MMWR)*, March 26, 1999. That most people labor under the delusion that restaurants are a more common source of bacterial illnesses than home kitchens is demonstrated in a survey published by S.B. Fein et al., "Foodborne illness: Perceptions, experience, and preventive behaviors in the United States." *Journal of Food Protection*. 1995; 58: 1405–11. Charles Gerba appeared in an interview on the CBS TV show *48 Hours* on September 28, 2000, where he said, according to the transcript provided by CBS News, "There are actually 200 times more fecal bacteria on the average cutting board in the home

than [on] the toilet seat." The Food and Drug Administration publishes several useful and informative pamphlets on eggs and egg-containing foods, which are available on their web site, www.fda.gov. The story of the 47-year-old man who ate some really bad beef stew is in the *MMWR* for March 24, 1995, and the details of the Alaskan problem can be found in the *MMWR* for August 17, 2001. The statistics on where you are likely to get botulism, and from which foods, are from Centers for Disease Control and Prevention: Botulism in the United States, 1899–1996. Handbook for Epidemiologists, Clinicians, and Laboratory Workers, 1998. *E. coli* and its appearance at the Washington County Fair are described in the *MMWR*, September 17, 1999. The shocking news that fast food restaurants are less contaminated with *E. coli* than fancier table service establishments was established by the 1997 FoodNet *E. coli* O157 case-control study carried out by the CDC. For information about resistant strains of *Campylobacter jejuni* that appeared after the introduction of fluoroquinolone, see K.E. Smith, J.M. Besser, F. Leano, J. Bender, J. Wicklund, B. Johnson et al., "Fluoroquinolone-resistant *Campylobacter* isolated from humans and poultry in Minnesota" [abstract]. Program of the 1st International Conference on Emerging Infectious Diseases; Atlanta, Georgia; 1998 Mar 7–10. Atlanta: Centers for Disease Control and Prevention; 1998. On the treatment of *Campylobacter* infection with antibiotics, see S.F. Altedruse, N.J. Stern, P.I. Fields, and D.L. Swerdlow, "*Campylobacter jejuni:* An emerging food-borne pathogen." *Emerging Infectious Diseases.* 1999; 1: 28–35. The sad fact that most broiler chickens are infected with *Campylobacter* is USDA-approved information: Food Safety and Inspection Service, *Nationwide broiler chicken microbiolgic baseline data collection program, 1994–1995* (Washington, D.C.: United States Department of Agriculture, 1996). The tale of the people in Alaska who wish they hadn't gone to the company picnic to meet up with Norwalk-like virus is told in the *MMWR* for March 17, 2000, and the oyster lovers who got Norwalk-like virus from their favorite food is recounted in the *MMWR* for June 24, 1994. Scombroid poisoning from seafood is discussed in S.H. Arnold and W.D. Brown, "Histamine toxicity from fish products." *Adv in Food Res.* 1978; 23: 113–54. That histamines are stronger in rotting fish than they are in goldenrod is alluded to in S.L. Taylor, "Histamine food poisoning: Toxicology and clinical aspects." *Crit Rev Toxicol.* 1986; 17(2): 91–128. On the strains of *Vibrio vulnificus* that cause death, see J.K. Jackson, R.L. Murphree, and M.L. Tamplin, "Evidence that mortality from *Vibrio vulnificus* infection results from single strains among heterogeneous populations in shellfish." *J Clin Microbiol.* 1997 Aug; 35(8): 2098–2101. The first U.S. report of infection by *Cyclospora cayetanensis* is in P. Huang, J.T. Weber, D.M. Sosin, P.M. Griffin, E.G. Long, J.J. Murphy, F. Kocka, C. Peters et al., "The first reported outbreak of diarrheal illness associated with Cyclospora in the United States." *Ann Intern Med.* 1995 Sep 15; 123(6): 409–14. The results of the survey of the

habits of those thrill seekers who eat runny eggs, pink hamburgers, unpasteurized milk, and raw shellfish are written up in B. Shiferaw, P. Cieslak et al., "Prevalence of high-risk food consumption and food-handling practices among adults: A multistate survey, 1996–1997." The Foodnet Working Group. *Journal of Food Protection*. 2000 Nov; 63(11): 1538–43.

Chapter 3. Toilet Training: Washing Hands Is the Best Revenge

The hand-washing survey was conducted in August and September 2000 for the American Society for Microbiology by Wirthlin Worldwide, replicating a study done in 1996 that had similar results. The study showing that daily baths may be pleasant but don't do much for hygiene is A.A. Hartmann, "Daily bath and its effect on the normal human skin flora: Quantitative and qualitative investigations of the aerobic skin flora" [author's translation; article in German]. *Arch Dermatol Res.* 1979 June 25; 265(2): 153–64. That bathing patients with antibacterial soap doesn't do much more than bathing with ordinary soap and water to prevent postoperative wound infections is shown in G.A. Ayliffe, M.F. Noy, J.R. Babb, J.G. Davies, and J. Jackson, "A comparison of pre-operative bathing with chlorhexidine-detergent and non-medicated soap in the prevention of wound infection." *J Hosp Infect.* 1983 Sept.; 4(3): 237–44. Even surgical personnel are not encouraged to bathe before operations because of the results of several studies, including R. Speers, H. Bernard, F. O'Grady, and R.A. Shooter, "Increased dispersal of skin bacteria into the air after shower-baths." *Lancet.* 1965; 1: 478–83; and G.S. Hall, C.A. Mackintosh, and P.N. Hoffman, "The dispersal of bacteria and skin scales from the body after showering and after application of a skin lotion." *J Hyg (Camb).* 1986; 97: 289–98. The two studies of hand washing among nurses are J. Ojajarvi, P. Makela, and I. Rantsalo, "Failure of hand disinfection with frequent hand washing: A need for prolonged field studies." *J Hyg (Camb).* 1977; 79: 107–19; and E.L. Larson, C.A. Hughes, J.D. Pyrek, S.M. Sparks, E.U. Cagatay, and J.M. Bartkus, "Changes in bacterial flora associated with skin damage on hands of health care personnel." *Am J Infect Control.* 1998 Oct; 26(5): 513–21. The study whose exact methodology our sense of decorum prevents us from describing but that showed that *Salmonella* bacteria can remain in a bathroom for up to four weeks after everyone has stopped having diarrhea is J. Barker and S.F. Bloomfield, "Survival of Salmonella in bathrooms and toilets in domestic homes following salmonellosis." *J Appl Microbiol.* 2000 Jul;89(1):137–44. The Chilean study demonstrating the fecal-oral transmission of typhoid fever is M.L. Alvarez, J. Espinoza, G. Figueroa, M. Diaz, and F. Wurgaft, "Hygiene habits at the family level and typhoid." [Article in Spanish]. *Rev Med Chil.* 1991 Jan; 119(1): 84–91. The toilet-flushing exercise to determine the residual bacteria in toilet bowls was carried out by C.P. Gerba,

W. Craig, and J.L. Melnick, "Microbiological hazards of household toilets: Droplet production and the fate of residual organisms." *Appl Microbiol.* 1975 Aug; 30(2): 229–37. The Georgetown University study demonstrating that wet soap has more bacteria on it than dry soap is S.J. Brook and I. Brook, "Contamination of bar soaps in a household setting." *Microbios.* 1993; 76(306): 55–57. That kitchens are germier than bathrooms is shown in P. Rusin, P. Orosz-Coughlin, and C. Gerba, "Reduction of faecal coliform, coliform and heterotrophic plate count bacteria in the household kitchen and bathroom by disinfection with hypochlorite cleaners." *J Appl Microbiol.* 1998 Nov; 85(5): 819–28. The disgusting stuff you can find on your toothbrush is delineated in S.S. Taji, and A.H. Rogers, ADRF Trebitsch Scholarship. "The microbial contamination of toothbrushes: A pilot study." *Aust Dent J.* 1998 Apr; 43(2): 128–30. The information that *Porphyromonas gingivalis* can hide in the handle of an electric toothbrush appears in L. Bunetel, S. Tricot-Doleux, G. Agnani, and M. Bonnaure-Mallet, "*In vitro* evaluation of the retention of three species of pathogenic microorganisms by three different types of toothbrush." *Oral Microbiology and Immunology.* 2000: 15; 313–16.

Chapter 4. Whiter Whites and Brighter Colors: Healthy Laundry

The armpit study is C. Zeng, A.I. Spielman, B.R. Vowels, J. J. Leden, K. Biemann, and G. Preti, "A human axillary odorant is carried by apolipoprotein D." *Proc Natl. Acad. Sci. USA.* 1996; 93: 6626–30. Yes, other parts of the body may also produce foul odors because of the presence of microorganisms. For a scientific discussion of one of them, see R. Stanek, D.D. Glover, S. Mya, and B. Larsen, "Evaluation of vaginal malodor and efficacy of treatment by high performance ion exchange chromatography." *Biomed Chromatogr.* 1987; 2(2): 76–78. There are at least five reports suggesting that airborne laundry lint might be the source of disease transmission: Centers for Disease Control. "Outbreak of viral hepatitis in the staff of a pediatric ward—California." *Morbidity and Mortality Weekly Report.* 1977; 28: 77–79; P.C. Shah, S. Krajden, J. Kane, and R.C. Summerbell, "Tinea corporis caused by *Microsporum canis*: Report of a nosocomial outbreak." *Eur J Epidemiol.* 1988; 4: 33–37; D. Barrie, P.N. Hoffman, J.A. Wilson, and J.H. Kramer, "Contamination of hospital linen by *Bacillus cereus*." *Epidemiol Infect.* 1994; 113: 297–306; S.M. Standaert, R.H. Hutcheson, and W. Schaffner, "Nosocomial transmission of *Salmonella* gastroenteritis to laundry workers in a nursing home." *Infect Control Hosp Epidemiol.* 1994; 15: 22–26; J. Pasternak, R. Richtmann, A.P.P. Ganme et al. "Scabies epidemic: Price and prejudice." *Infect Control Hosp Epidemiol.* 1994; 15: 540–542. As for medical students' little white coats becoming contaminated, see W. Loh, V.V. Ng, and J. Holton, "Bacterial flora on the white coats of medical students." *J Hosp Infect.* 2000; 45: 65–68. The rules about hospital

laundry can be found in *Guidelines for Laundry in Health Care Facilities*, published by the Office of Health and Safety, Centers for Disease Control and Prevention, updated June 20, 2001. The study showing that washing surgical scrubs at home is just as effective as having them washed in a hospital laundry is N.L. Belkin, "Home laundering of soiled surgical scrubs: Surgical site infections and the home environment." *Am J Infect Control.* 2001 Feb; 29(1): 58–64. Dry cleaning makes clothes look good; it doesn't make them sterile: see C.J. Bates, M.H. Wilcox, T.L. Smith, and R.C. Spencer, "The efficacy of a hospital dry cleaning cycle in disinfecting material contaminated with bacteria and viruses." *J Hosp Infect.* 1993; 23: 255–262.

Chapter 5. Clean Up That Room: Kids and Microbes

The information on vaccination is from the CDC's Advisory Committee on Immunization Practices, whose comprehensive recommendations and recommendations for specific diseases are available on their web site at www.cdc.gov/nip/publications/ACIP-list.htm. These recommendations, of course, change from time to time, and they are updated regularly. The supposed association between inflammatory bowel disease and the MMR vaccine is refuted in R.L. Davis, P. Kramarz et al. "Measles-mumps-rubella and other measles-containing vaccines do not increase the risk for inflammatory bowel disease: A case-control study from the Vaccine Safety Datalink project." *Arch Pediatr Adolesc Med.* 2001 Mar; 155(3): 354–9. For a refutation of the supposed association between autism and the MMR vaccine, see S. DeWilde, I.M. Carey, N. Richards, S.R. Hilton, and D.G. Cook, "Do children who become autistic consult more often after MMR vaccination?" *Br J Gen Pract.* 2001 Mar; 51(464): 226–7. The risks of allowing even as few as 2 percent of children to forego measles and pertussis vaccination are elucidated in D.R. Feikin, D.C. Lezotte, R.F. Hamman, D.A. Salmon, R.T. Chen, and R.E. Hoffman, "Individual and community risks of measles and pertussis associated with personal exemptions to immunization." *JAMA.* 2000 Dec 27; 284(24): 3145–50. The original recommendation for the use of rotavirus vaccine was published in the *Morbidity and Mortality Weekly Report* for March 19, 1999, vol. 48, no. RR-2, and then withdrawn in the issue published November 5, 1999, vol. 48, no. 43. The possibility that bottle feeding may be a risk for rotavirus infection appears in D. Sethi et al., "A study of infectious intestinal disease in England: Risk factors associated with group A rotavirus in children." *Epidemiol Infect.* 2001 Feb; 126(1): 63–70. The information about transmission of shigellosis in day care centers is in J.C. Mohle-Boetani, M. Stapleton, R. Finger, N.H. Bean, J. Poundstone, P.A. Blake, and P.M. Griffin, "Communitywide shigellosis: Control of an outbreak and risk factors in child day-care centers." *Am J Public Health.* 1995 June; 85(6): 812–6. On using ribavirin to treat respiratory syn-

cytial virus infection, see S.E. Ohmit, F.W. Moler, A.S. Monto, and A.S. Khan, "Ribavirin utilization and clinical effectiveness in children hospitalized with respiratory syncytial virus infection." *J Clin Epidemiol.* 1996 Sept.; 49(9): 963–67. The question of using Synagis for prevention of RSV is also discussed in S.E. Ohmit, F.W. Moler, A.S. Monto, and A.S. Khan, "Ribavirin utilization and clinical effectiveness in children hospitalized with respiratory syncytial virus infection." *J Clin Epidemiol.* 1996 Sept; 49(9): 963–67 and in F. Moler, "RSV immune globulin prophylaxis: Is an ounce of prevention worth a pound of cure?" *Pediatrics.* 1999 Sept; 104(3 Pt 1): 559–60. There is also a substantial article by Melody Petersen on the subject in the *New York Times* of January 31, 2001, section C, page 1. For an account of methicillin-resistant *Staphylococcus aureus* in U.S. day care centers, see P.M. Adcock, P. Pastor, F. Medley, J.E. Patterson, and T.V. Murphy, "Methicillin-resistant *Staphylococcus aureus* in two child care centers." *J Infect Dis.* 1998 Aug; 178(2): 577–80. The accounts of the four children infected with methicillin-resistant *Staphylococcus aureus* appeared in the *MMWR* for August 20, 1999. The Danish study showing that kids in day care get more antibiotics than kids who stay at home is N. Thrane, C. Olesen, J.T. Md, C. Sondergaard, H.C. Schonheyder, and H.T. Sorensen, "Influence of day care attendance on the use of systemic antibiotics in 0- to 2-year-old children." *Pediatrics.* 2001 May; 107(5): E76. The resistance of *Prpriionobacterium acnes* to antibiotics is reviewed in J. Cooper, "Systematic review of *Propionibacterium acnes* resistance to systemic antibiotics." *Med J Aust.* 1998 Sep 7; 169(5): 259–61. The meningitis outbreak in Alliance, Ohio, was reported in, among many other places, the *New York Times* of June 6, 2001, section A, page 18. Additional information about vaccines, vaccine supply, and contraindications for immunization is available at http://www.cdc.gov/nip or at the National Immunization hotline: (800) 232-2522 (English), or (800) 232-0233 (Spanish). Copies of the schedule can be obtained at http://www.cdc.gov/nip/recs/child-schedule.htm. Approved by the Advisory Committee on Immunization Practices (http://www.cdc.gov/nip/acip), the American Academy of Pediatrics (http://www.aap.org), and the American Academy of Family Physicians (http://www.aafp.org).

Chapter 6. What Love's Got to Do with It: Microbes and Your Sex Life

The account of the Minnesota wrestling camp can be found in the *Morbidity and Mortality Weekly Report* for February 9, 1990. Alan M. Kraut, *Silent Travelers: Germs, Genes, and the "Immigrant Menace"* (Baltimore: Johns Hopkins University Press, 1994) details the long history of blaming immigrants for disease, a habit that persists in America even today. The Rothschilds' work on evidence of syphilis in human skeletons is described in *Archaeology*, January–February 1997. The account of ciprofloxacin- resistant gonorrhea in Hawaii

appears in the *MMWR* for September 22, 2000. The information about the epidemiology of human papillomavirus is from W. Cates Jr., "Estimates of the incidence and prevalence of sexually transmitted diseases in the United States." American Social Health Association Panel. *Sex Transm Dis.* 1999 Apr; 26(4 Suppl): S2–7 and from L. Koutsky, "Epidemiology of genital human papillomavirus infection." *Am J Med.* 1997 May 5; 102(5A): 3–8. The information about young women and HPV is in the following articles: G.Y. Ho, R. Bierman, L. Beardsley, C.J. Chang, and R.D. Burk, "Natural history of cervicovaginal papillomavirus infection in young women." *N Engl J Med.* 1998 Feb 12; 338(7): 423–28; R.D. Burk, G.Y. Ho, L. Beardsley, M. Lempa, M. Peters, and R. Bierman, "Sexual behavior and partner characteristics are the predominant risk factors for genital human papillomavirus infection in young women." *J Infect Dis.* 1996 Oct; 174(4): 679–89; and H.M. Bauer, Y. Ting, C.E. Greer, J.C. Chambers et al., "Genital human papillomavirus infection in female university students as determined by a PCR-based method." *JAMA.* 1991 Jan 23–30; 265(4): 472–77. The report on the increase in pertussis incidence among infants is from the *MMWR* of February 1, 2002. There are several diseases that are now considered candidates for eradication, including measles, rubella, yellow fever, leprosy, and *Haemophilus* influenza type B (Hib), among others. For a good discussion of the feasibility of eradication, see the unnumbered supplement to the *MMWR* of December 31, 1999, "Global Disease Elimination and Eradication as Public Health Strategies."

Chapter 7. Wild Kingdom: Pets and Their Germs

All of these cases of reptile-associated salmonellosis are described in the *Morbidity and Mortality Weekly Report* for November 12, 1999. The cases associated with birds appear in the *MMWR* for April 14, 2000. There are a number of studies that suggest that companion animals are good for you. Here are three examples: E.L. Altschuler, "Pet-facilitated therapy for posttraumatic stress disorder." *Ann Clin Psychiatry.* 1999 Mar; 11(1): 29–30; S.L. Nagengast, M.M. Baun, M. Megel, and J.M. Leibowitz, "The effects of the presence of a companion animal on physiological arousal and behavioral distress in children during a physical examination." *J Pediatr Nurs.* 1997 Dec; 12(6): 323–30; and J. Jorgenson, "Therapeutic use of companion animals in health care." *Image J Nurs Sch.* 1997; 29(3): 249–54. The frightening story of the kids and the kittens is told in the *MMWR* for December 16, 1994. For the technical details on cat scratch disease, see J.J. Windsor, "Cat-scratch disease: Epidemiology, aetiology and treatment." *Br J Biomed Sci.* 2001; 58(2): 101–10. The discovery of the connection between *B. henselae* and cat scratch disease is alluded to in B.B. Chomel, "Cat-scratch disease." *Rev Sci Tech.* 2000 Apr; 19(1): 136–50. Swenson's Wild Midwest Exotic Petting Zoo is immortalized in *MMWR*,

September 3, 1999. The account of the pregnant shepherdess appears in *Emerging Infectious Disease*, April-June 1997. Information on the epidemiology of *Campylobacter* infection is in R.V. Tauxe, "Epidemiology of *Campylobacter jejuni* infections in the United States and other industrialized nations." I. Nachamkin, M.J. Blaser, and Tompkins, eds., *Campylobacter jejuni: Current Status and Future Trends* (Washington, D.C.: American Society for Microbiology, 1992), 9–19.

Chapter 8. Up Your Nose: The Flu and the Cold

There are good two books on the 1918 flu. The first is Alfred W. Crosby, *America's Forgotten Pandemic: The Influenza of 1918* (New York: Cambridge University Press, 1989). The other, which adds details about the search to identify the 1918 flu virus and relies heavily on Crosby, is Gina Kolata's *Flu: The Story of the Great Influenza Pandemic of 1918 and the Search for the Virus That Caused It* (New York: Farrar Straus & Giroux, 1999). The experiments with mice and the 1918 virus are described in C.F. Basler et al., "Sequence of the 1918 pandemic influenza virus nonstructural gene (NS) segment and characterization of recombinant viruses bearing the 1918 NS genes." *Proc Natl Acad Sci U S A*. 2001 Feb 27; 98(5): 2746–51. The first genetic description of the 1918 flu is J.K. Taubenberger, A.H. Reid, A.E. Krafft, K.E. Bijwaard, and T.G. Fanning, "Initial genetic characterization of the 1918 'Spanish' influenza virus." *Science*. 1997 Mar 21; 275(5307): 1793–96. The survey of Medicaid patients who were wrongly prescribed antibiotics is reported in A.G. Mainous, 3rd, W.J. Hueston, and J.R. Clark, "Antibiotics and upper respiratory infection: Do some folks think there is a cure for the common cold." *J Fam Pract*. 1996 Apr; 42(4): 357–61. The completed viral gene sequences of the 1918 flu virus appear in J.K. Taubenberger, A.H. Reid, and T.G. Fanning, "The 1918 influenza virus: A killer comes into view." *Virology*. 2000 Sept 1; 274(2): 241–45. For a discussion of why there are winter and summer diseases, see S.F. Dowell, "Seasonal variation in host susceptibility and cycles of certain infectious diseases." *Emerg Infect Dis*. 2001 May-June; 7(3): 369–74. On the effectiveness of the hand-to-hand transmission of colds, see J.M. Gwaltney Jr., P.B. Moskalski, and J.O. Hendley, "Hand-to-hand transmission of rhinovirus colds." *Ann Intern Med*. 1978 Apr; 88(4): 463–67.

Chapter 9. Bottled or Tap: Water, and What's in It

The unplanned and unpleasant interactions between visitors and *E. coli* at the Washington County Fair in 1999 are discussed in the *Morbidity and Mortality Weekly Report* for September 17, 1999. The Milwaukee outbreak of cryptosporidium is described in W.R. MacKenzie et al., "A massive outbreak in

Milwaukee of *Cryptosporidium* infection transmitted through the public water supply," *N. Engl. J. Med.* 1994; 331: 161–67. The false alarm in Washington, D.C., is reported in the *MMWR* for September 16, 1994. The only outbreak of cryptosporidiosis traced to bottled water in areas controlled by the United States is outlined in the *MMWR* dated April 12, 1996, vol. 45, no. SS-1. Some of the information about brands of bottled water comes from *The Good Water Guide* by Maureen and Timothy Green, rev. ed. (London: Rosendale Press, 1994). Some of it comes right off the labels of the bottled water itself—a very reliable source of information, thanks to the regulations of the Food and Drug Administration. The outbreaks of *E. coli* in Wyoming, Illinois, and Washington State and the shigella infection in Minnesota are all described in the *MMWR* for May 26, 2000, vol. 49, no. SS-4. This is the latest surveillance report for waterborne disease outbreaks.

Chapter 10. Fresh Air and Sunshine: Outdoor Fun with Microbes

The case of malaria in the 11-year-old boy in Suffolk County, New York, appears in the *Morbidity and Mortality Weekly Report* for June 9, 2000. The reasons why researchers now believe that malaria is a relatively new disease appear in S.A. Tishkoff et al., "Haplotype diversity and linkage disequilibrium at human G6PD: Recent origin of alleles that confer malarial resistance." *Science.* 2001 Jul 20; 293(5529): 455–62. Recent outbreaks of malaria are discussed in J.R. Zucker, "Changing patterns of autochthonous malaria transmission in the United States: A review of recent outbreaks." *Emerg Infect Dis.* 1996 Jan-Mar; 2(1): 37–43. The Westchester County, New York, study of Lyme disease treatment is R.B. Nadelman et al., "Prophylaxis with single-dose doxycycline for the prevention of Lyme disease after an *Ixodes scapularis* tick bite." *N Engl J Med.* 2001 Jul 12; 345(2): 79–84. The serosurvey of West Nile virus is unpublished data from the CDC and the New York City Department of Health. The quotation from the *New York Times* about Lyme disease anxiety appeared in an article by Gina Kolata on June 13, 2001, section A, page 1. The account of the man from Kennebec County, Maine, with Powassan encephalitis appears in the *MMWR* for September 7, 2001. The shigella outbreak in the water park in Volusia County, Florida, is reported in the *MMWR* for June 30, 2000.

Chapter 11. Paint the Town Red: Germs in Public Places

The case of the Indian boy whose gums were infested with fly larvae is detailed in A.P. Bhatt and A. Jayakrishnan, "Oral myiasis: A case report." *Int J. Paediatr Dent.* 2000 Mar; 10(1): 67–90. The idea that electrocuting insect traps might spread disease is discussed in J.E. Urban and A. Broce, "Killing of flies in electrocuting insect traps releases bacteria and viruses." *Curr Microbiol.*

2000 Oct; 41(4): 267–70. To give proper credit for the vivid phrase "refuse and promiscuous-landing synanthropic filth flies": T.K. Graczyk, M.R. Cranfield, R. Fayer, and H. Bixler, "House flies (*Musca domestica*) as transport hosts of *Cryptosporidium parvum.*" *Am J Trop Med Hyg.* 1999 Sept; 61(3): 500–504. The deadly *E. coli* O157:H7 found in flies on a Japanese farm is reported in M. Iwasa, S. Makino, H. Asakura, H. Kobori, and Y. Morimoto, "Detection of *Escherichia coli* O157:H7 from *Musca domestica* (Diptera: Muscidae) at a cattle farm in Japan." *J Med Entomol.* 1999 Jan; 36(1): 108–12. The good news that *H. pylori* can't be transmitted by houseflies appears in M.S. Osato, K. Ayub, H.H. Le, R. Reddy, and D.Y. Graham, "Houseflies are an unlikely reservoir or vector for *Helicobacter pylori.*" *J Clin Microbiol.* 1998 Sept; 36(9): 2786–88. That the quality of the body odor depends on the quantity of bacteria was revealed by J.J. Leyden, K.J. McGinley, E. Holzle, J.N. Labows, and A.M. Kligman, "The microbiology of the human axilla and its relationship to axillary odor." *J Invest Dermatol.* 1981 Nov; 77(5): 413–16. The report on the outbreak of *M. fortuitum* in nail salons in California and the quotation from Dr. Winthrop appeared in the *New York Times* on April 27, 2001 in an article by Lawrence Altman, section A, page 22. Dr. Aaron Glatt's unscientific survey of the New York City subway system is reported in the *Times* of January 30, 2001, in an article by Randy Kennedy, section B, page 8. That *Salmonella* can persist for some time after an infected person flushes the toilet and that the flushing itself can cause contamination of the toilet seat and lid is shown in J. Barker and S.F. Bloomfield, "Survival of *Salmonella* in bathrooms and toilets in domestic homes following salmonellosis." *J Appl Microbiol.* 2000 Jul; 89(1): 137–44. Spreading Norwalk-like virus in shared college bathrooms was studied in C.L. Moe, W.A. Christmas, L.J. Echols, and S.E. Miller, "Outbreaks of acute gastroenteritis associated with Norwalk-like viruses in campus settings." *J Am Coll Health.* 2001 Sept; 50(2): 57–66, and the interesting information about dirty bathrooms versus dirty hands as a source of infection was presented in E.E. Ekanem, H.L. DuPont, L.K. Pickering, B.J. Selwyn, and C.M. Hawkins, "Transmission dynamics of enteric bacteria in day-care centers." *Am J Epidemiol.* 1983 Oct; 118(4): 562–72. Legionnaire's disease on the cruise ship *Horizon* was reported in the *Morbidity and Mortality Weekly Report* for August 12, 1994, and the report on the *Shigella flexneri* outbreak on the *Viking Serenade* appears in the *MMWR* for September 9, 1994. That bacterial counts in airplanes were somewhat lower than those in public buildings was reported in C. Thibeault, "Cabin air quality." Passenger Health Subcommittee, Air Transport Medicine Committee, Aerospace Medical Association. *Aviation Space Environ Med.* 1997; 68(1): 80–82, and R.B. Rayman, "Passenger safety, health, and comfort: A review." *Aviation Space Environ Med.* 1997; 68(5): 432–40. The six instances of passengers or crew who flew while infected with TB are outlined in the *MMWR* dated March 3, 1995. The flu outbreak on the

plane confined to the runway for three hours is reported in M.R. Moser, T.R. Bender, H.S. Margolis, G.R. Noble, A.P. Kendal, and D.G. Ritter, "An outbreak of influenza aboard a commercial airliner." *Am J Epidemiol.* 1979 Jul; 110(1): 1–6. The malaria cases on the flight from Lebanon to São Paulo are reported in S. Cimerman et al., "Malaria transmission associated with airplane travel." *Braz J Infect Dis.* 1997 Jun; 1(3): 135–37. On tattoos and their non-association with the risk for hepatitis, see A.L. Silverman, J.S. Sekhon, S.J. Saginaw, D. Wiedbrauk, M. Balasubramaniam, and S.C. Gordon, "Tattoo application is not associated with an increased risk for chronic viral hepatitis." *Am J Gastroenterol.* 2000 May; 95(5): 1312–15.

Chapter 12. The Antiseptic Supermarket: Products That Do Something, Products That Do Nothing, and Products That Actually Do Harm

For a refutation of Mr. Cueman's assertion that there is no evidence of a link between resistant bacteria and triclosan, see, for example, the following two articles: R.J. Heath, J. Li, G.E. Roland, C.O. Rock, "Inhibition of the *Staphylococcus aureus* NADPH-dependent enoyl-acyl carrier protein reductase by triclosan and hexachlorophene." *J Biol Chem.* 2000 Feb 18; 275(7): 4654–59, and L.M. McMurry, M. Oethinger, and S.B. Levy, "Overexpression of marA, soxS, or acrAB produces resistance to triclosan in laboratory and clinical strains of *Escherichia coli.*" *FEMS Microbiol Lett.* 1998 Sep 15; 166(2): 305–9. The two studies that show that using antibacterial hand cleansers is better than using nothing at all were both sponsored by companies that make antibacterial hand cleansers: D.L. Dyer, A. Shinder, and F. Shinder, "Alcohol-free instant hand sanitizer reduces elementary school illness absenteeism." *Fam Med.* 2000 Oct; 32(9): 633–38. The authors are employed by Woodward Laboratories, which manufactures several antibacterial hand sanitizers, including one called HandClens. Also, B. Hammond, Y. Ali, E. Fendler, M. Dolan, and S. Donovan, "Effect of hand sanitizer use on elementary school absenteeism." *Am J Infect Control.* 2000 Oct; 28(5): 340–46. These authors are employed by GOJO Industries, manufacturers of Purell hand cleaner, an alcohol-based product. Neither study compares hand sanitizers to ordinary soap and water. For that comparison, and a showing that ordinary soap and water may actually be better than antibacterial hand cleansers, see D.L. Charbonneau, J.M. Ponte, and B.A. Kochanowski, "A method of assessing the efficacy of hand sanitizers: Use of real soil encountered in the food service industry." *J Food Prot.* 2000 Apr; 63(4): 495–501. The authors are employees of the Procter & Gamble Co., which manufactures Ivory Soap, among many other cleaning products, but no hand sanitizers. P & G does make an antibacterial dishwashing liquid called Dawn, which the company says can be used as an

antibacterial hand soap. The survey on the supermarket availability of antibacterial agents appears in E.N. Perencevich, M.T. Wong, and A.D. Harris, "National and regional assessment of the antibacterial soap market: A step toward determining the impact of prevalent antibacterial soaps." *Am J Infect Control.* 2001 Oct; 29(5): 281–83.

In addition to the books mentioned above, the following provided useful background information:

Biddle, Wayne. *A Field Guide to Germs.* New York: Henry Holt, 1995.

Morse, Stephen S., ed. *Emerging Viruses.* New York: Oxford University Press, 1993.

Scott, Elizabeth, and Paul Sockett. *How to Prevent Food Poisoning.* New York: Wiley, 1998.

Tomes, Nancy. *The Gospel of Germs: Men, Women, and the Microbe in American Life.* Cambridge: Harvard University Press, 1998.

Index

Printed in the USA
CPSIA information can be obtained
at www.ICGtesting.com
JSHW012014140824
68134JS00025B/2416